SpringerBriefs in Applied Sciences
and Technology

SpringerBriefs in Computational Intelligence

Series Editor

Janusz Kacprzyk, Systems Research Institute, Polish Academy of Sciences,
Warsaw, Poland

SpringerBriefs in Computational Intelligence are a series of slim high-quality publications encompassing the entire spectrum of Computational Intelligence. Featuring compact volumes of 50 to 125 pages (approximately 20,000-45,000 words), Briefs are shorter than a conventional book but longer than a journal article. Thus Briefs serve as timely, concise tools for students, researchers, and professionals.

More information about this subseries at http://www.springer.com/series/10618

Tiago Martins · Rui Neves

Stock Exchange Trading Using Grid Pattern Optimized by A Genetic Algorithm with Speciation

The Case of S&P 500

Springer

Tiago Martins
Instituto Superior Técnico
Instituto de Telecomunicações
Lisbon, Portugal

Rui Neves
Instituto Superior Técnico
Instituto de Telecomunicações
Lisbon, Portugal

ISSN 2191-530X   ISSN 2191-5318 (electronic)
SpringerBriefs in Applied Sciences and Technology
ISSN 2625-3704   ISSN 2625-3712 (electronic)
SpringerBriefs in Computational Intelligence
ISBN 978-3-030-76679-5   ISBN 978-3-030-76680-1 (eBook)
https://doi.org/10.1007/978-3-030-76680-1

This Springer imprint is published by the registered company Springer Nature Switzerland AG
The registered company address is: Gewerbestrasse 11, 6330 Cham, Switzerland

Tiago Mousinho Martins
To my partents, grandparents and Margarida

Rui Ferreira Neves
To Susana and Tiago

Preface

Financial markets have an almost random behavior. To be successful in trading a system needs to adapt constantly. To achieve better results in this book a genetic algorithm optimizes a grid template pattern detector to find the best point to trade in the SP 500. The pattern detector is based on a template using a grid of weights with a fixed size. The template takes in consideration not only the closing price but also the open, high and low values of the price during the period under testing in contrast to the traditional methods of analyzing only the closing price. Each cell of the grid encompasses a score, and these are optimized by an Evolutionary Genetic Algorithm that takes genetic diversity into consideration through a speciation routine, giving time for each individual of the population to be optimized within its own niche. With this method the system is able to present better results and improve the results compared with other template approaches. It was tested considering real data from the stock market and against state-of-the-art solutions, namely, the ones using a grid of weights which does not have a fixed size and non speciated approaches. During the testing period the presented solution had a return of 21.3% compared to 10.9% of the existing approaches. The use of speciation was able to increase the returns of some results as genetic diversity was taken into consideration.

Lisbon, Portugal
March 2021

Tiago Martins
Rui Neves

Contents

Acronyms

ANN	Artificial Neural Networks
API	Application Programming Interface
ATR	Average True Range
DEAP	Distributed Evolutionary Algorithms in Python
EMA	Exponential Moving Average
FSGW	Fixed Size Grid of Weights
GA	Genetic Algorithm
HMA	Hull Moving Average
MACD	Moving Average Convergence Divergence
NE	NeuroEvolution
NEAT	NeuroEvolution of Augmenting Topologies
OBV	On Balance Volume
PIPs	Perceptually Important Points
ROI	Return on Investment
RSI	Relative Strength Index
SMA	Simple Moving Average
SMTP	Simple Mail Transfer Protocol
TSAG	Traditional Simple Adaptable Grid
VIX	Chicago Board Options Exchange Volatility Index

List of Figures

List of Tables

Chapter 1
Introduction

The domain of computational finance has been earning increased attention by people from both computational intelligence and finance domains. This happens because, when studying the market, determining the optimal timings to buy or sell stocks is crucial in achieving a trading strategy that will accomplish the financial performance requirements of the investor. However, this process tends to be very difficult as the stock market is noisy, non-stationary and non-linear, and at the same time influenced by many other factors, such as political, economic and even psychological variables.

1.1 Background

The efficient market hypothesis [2] presents a view in which is not possible beat the market because it already includes all available information at a specific moment, as well as defends that stock prices behave in a random walk, being the prediction of future prices impossible as they cannot be reflected by any trend or pattern. Besides that, it is also stated that the passive investing strategy Buy and Hold (B&H), where an investor buys stocks and holds them for a long amount of time, despite of market fluctuations, is the more adequate strategy.

Adversely to this vision, some investigators support the thesis that markets are not efficient, existing some variables that they do not reflect right away, which opens a window of opportunity to beat them and obtain returns using stock market forecasting techniques, such as technical analysis, fundamental analysis, and analysis of time series. To achieve this, investment strategies that are able to process large amounts of data and generate appropriate trading signals (buy or sell) must be defined, being the main challenges of the computational finance domain trying to predict future trend prices, in order to obtain the best correlation between risk (that should be as low as possible) and profit (that should be as high as possible).

© The Author(s), under exclusive license to Springer Nature Switzerland AG 2021   1
T. Martins and R. Neves, *Stock Exchange Trading Using Grid Pattern Optimized by A Genetic Algorithm with Speciation*, SpringerBriefs in Computational Intelligence,
https://doi.org/10.1007/978-3-030-76680-1_1

One way of predicting the behavior of the market used by traders is through the analysis and identification of chart patterns in historical prices of financial assets. A problem arises with this approach regarding the way the market should match a specific chart pattern to make a buy or sell decision. Several works like [1, 3, 5, 6] have been published where different approaches to deal with the chart pattern identification are proposed and tested.

1.2 Proposed Solution

In this report a new approach to pattern identification is presented, combining the template based approach using a grid of weights, proposed by [5], and a GA [4]. The grid of weights assigned for a specific template and placed above the signal is going to generate a score that will be higher as close as the signal is from the pattern. The GA will be used to optimize the weights assigned to each cell of the grid so the pattern identification can be more precise, avoiding false positives and false negatives.

1.3 Main Contribution

The main contributions of the system proposed in this report are:

1. The creation of a new way to attribute the score to the signal, in which not only the closing price is taken into consideration, but also the price fluctuations over each time interval of the time series, in order to retrieve more information from it and improve the accuracy of the algorithm's decision;
2. The fact that grid scores are not fixed but instead optimized by the GA in order to find the most profitable solution;
3. The elimination of the adaptability of the grid imposed by the previous approach, in which the range of prices of the sliding window would decide how the size of the grid would vary, into a scenario where fixed dimensions are attributed to the grid prior to its use in order to ensure that all detected patterns are identical, not only to the template, but also within each other;
4. The preservation of genetic diversity in the population being optimized by the GA, through the use of a speciation method, in order to give time for different genotypic solutions to be optimized within their own niche.

1.4 Outline

This document describes the research and work developed and it is organized as follows:

- In this chapter presents the motivation, background and proposed solution;
- Chapter 2 describes the previous work in the field;
- Chapter 3 describes the system architecture and its requirements;
- Chapter 4 describes the evaluation tests performed and the corresponding results;
- Chapter 5 summarizes the work developed and future work.

References

1. Cervelló-Royo, R., Guijarro, F., Michniuk, K.: Stock market trading rule based on pattern recognition and technical analysis: forecasting the DJIA index with intraday data. Expert Syst. Appl. **42**, 5963–5975 (2015)
2. Fama, E.F.: Efficient capital markets: a review of theory and empirical work. J. Financ. **25**(2), 383–417 (1970)
3. Fu, T., Chung, F., Luk, R., Ng, C.: Stock time series pattern matching: template-based vs rule-based approaches. Eng. Appl. Artif. Intell. **20**, 347–364 (2007)
4. Holland, J.H.: Adaptation in Natural and Artificial Systems: An Introductory Analysis with Applications. The MIT Press, Cambridge (1992)
5. Leigh, W., Paz, N., Russell, P.: Market timing: a test of a charting heuristic. Econ. Lett. **77**, 55–63 (2002)
6. Leitão, J., Neves, R., Horta, N.: Combining rules between PIPs and SAX to identify patterns in financial markets. Expert Syst. Appl. 242–254 (2016)

Chapter 2
Related Work

2.1 Financial Portfolio and Portfolio Optimization

A financial portfolio consists in allocating wealth among several financial assets, also called securities or investments, i.e. stocks. The portfolio optimization problem is one of the most important research themes in recent risk management studies, especially regarding its two most relevant parameters: risk, which is intended to be minimized as much as possible, and return, that should be as large as possible. Nevertheless, these two parameters are not inversely proportional as we would wish, but instead a high return is always associated with higher risk [7, 16].

2.2 Stock Market Analysis

Regarding stock market analysis, two main approaches arise: fundamental analysis and technical analysis. Fundamental analysis takes advantage of available data about companies to understand which ones are on the same competitive industry and measure the intrinsic value of a stock, in order to take conclusions on which is the best to invest, based on ratios that measure profitability, liquidity, debt and growth. Technical analysis does not attempt to measure the intrinsic value of a stock, but instead uses charts with historical data to identify trends and patterns and tries to predict what the stock may do in the future. This is done through the use of technical indicators that will define the behavior of the program [28].

© The Author(s), under exclusive license to Springer Nature Switzerland AG 2021   5
T. Martins and R. Neves, *Stock Exchange Trading Using Grid Pattern Optimized by A Genetic Algorithm with Speciation*, SpringerBriefs in Computational Intelligence, https://doi.org/10.1007/978-3-030-76680-1_2

2.2.1 Technical Indicators

A technical indicator is a metric usually extracted from the price of an asset and its main goal is to help predicting future prices or incoming trends. Some of the most used technical indicators are presented below [16, 23].

2.2.1.1 Trend Following Indicators

This type of indicators have the main goal of identify shifts in the tendency, namely if a trend has begun or finishing its cycle.

- Simple and Exponential Moving Average—The SMA is one of the oldest technical indicators, used to find the mean price over a time window. It is calculated simply by averaging the price data over a defined time interval in order to produce a smooth line for an easier analysis, instead of trying to understand the irregular signal of prices. The EMA, works in the same way as the SMA, but assigns more weight to the most recent data, thus making this average more sensitive to it. For this reason, EMA is much more used than SMA.
- Hull Moving Average—EMA's major problem is related to the buying and selling signals that it generates, as there is a lag between the time an investor should be making a move on the stock and the actual signaling from the average. The HMA fights that problem defining an even smoother curve that follow the price much more closely than its predecessor, making the generation of signals much more precise.
- Double Crossover—The Double Crossover method consists on combining two different moving averages to generate buy and sell signals i.e. each time they cross each other. It is usually used both a short and long moving averages to take advantage of their characteristics: the shorter is more sensible to the market signal and reacts faster, but can produce false signals; the longer has more delay but produces more accurate results.

2.2.1.2 Momentum Oscillators and Indicators

These indicators are important to identify turning points and are useful in markets that are not biased i.e. volatile and hardly predictable. The moving averages also identify and follow trends, but using them on these agitated markets might generate false signals.

- Relative Strength Index—The RSI is a momentum oscillator used to measure the stock's recent gains against its recent losses and tries to identify overbought or oversold conditions. When calculated, the RSI signal has a range from 0 to 100 and there are specific levels that, if crossed by the RSI line, must generate buy or sell signals.

- Moving Average Convergence Divergence—MACD is a trend following momentum indicator representing the relation between two moving averages. It is stated as one of the most reliable indicators and is defined by two lines: the MACD line (corresponding to the difference between a 12-week EMA and a 26-week EMA) and a trigger line (corresponding to a 9-day EMA applied to the MACD values). A histogram is obtained from the difference between the former lines, based on which buy and sell signals are generated.
- On Balance Volume—The OBV is a momentum indicator that tries to relate the way volume is flowing in a security with the price, assuming that, if price changes alongside volume we are facing a tendency, otherwise it must be an indication that the tendency is about to finish.

2.2.1.3 Volatility Indicators

Volatility indicators are a reflection of investor sentiment and predictability of signal price. A stock is considered volatile if its signal movements are instable and with constant changes in its amplitude. In this kind of stocks, the risk associated tends to be higher and the probability of having a big success or a big failure in the investment increases. On the other hand, a low volatile stock is characterized by a stable and predictable signal, reducing the probability of having a big success or failure, which reflects in lower risk for the investment.

- COBE Volatility Index—The VIX is commonly introduced as the "fear index" and measures the implied volatility of S&P 500. The VIX is quoted in percentage points and produces the expected range of movement in the previous index over the next year.
- Average True Range—ATR was designed with commodities and daily prices in mind. These are more volatile than stocks and are subject to gaps and limit moves that would not be captured by volatility formulas based on high-low range. ATR is based on True Range which uses absolute price changes and reflects the enthusiasm behind a move or a breakout [33].

2.3 Optimization Techniques

2.3.1 Genetic Algorithms

In artificial intelligence we find the concept of Evolutionary Algorithms that use mechanisms inspired by biological evolution to find the best individuals in a population. One of the best known evolutionary techniques is the GA, which is based on Darwin's principle that the fittest survive [2, 17, 22].

GAs start with a randomly generated population where each individual is defined by a set of chromosomes (also called genome) with encoded data in its genes. In the case of portfolio optimization, this data usually represents financial ratio weights or trading parameters. Despite the fact that GAs are randomized, it is not considered a random procedure, as it is follows an iterative process inspired by biological evolution and constituted by several steps, such as evaluation, selection, crossover and mutation, in order to search solutions with good performance. In the evaluation stage, each individual is appraised through the use of an evaluation or fitness function, so the best ones are able to be within the population of the next generation, in a process called elitism selection, which simulates the "survival of the fittest" convention. Once the new generation is set the process continues with a crossover operation, where two previously chosen individuals, called parents, are merged to create a descendent with characteristics from both of them, with the purpose of combining their best attributes and generate a better performing individual. Finally a mutation is introduced to the chromosome in order to generate some diversity in the population, by randomly changing the genes in the individual. The GA process finishes when a termination condition is achieved, e.g. an established number of generations or a fitness value threshold.

Its capability of optimizing parameters in widely complex datasets makes it extremely indicated to be used in the financial markets to improve the investment strategy.

2.3.2 *Artificial Neural Networks*

ANN are computational models inspired by biological behaviour and adaptive capabilities of central nervous systems. GAs are used for supervised regression problems and are capable of solving non-linear problems with high complexity, being characterized for having a learning phase using existing datasets for predicting future results.

They are composed by interconnected processing units, called nodes or neurons, and they can be classified as input, output and hidden nodes. Input nodes receive the dataset of inputs of the model, output nodes play the role of outputting the predicted result, being their number usually the same as the classes to predict, and hidden nodes process information provided by input nodes, serving as an intermediary between input and output nodes, and their number is dependent of the task of the model [18].

The channels that connect neurons are called synapses and their job consists on taking a value from their input node, multiply it by a specific weight (called synaptic weight) and output the result to their output node. Each neuron has an activation function that is applied to the output of all its synapses and allows the neural network to introduce complex non-linear dynamic patterns into the model, like the ones we find in the stock markets.

In the learning phase, the backpropagation method [32] is used to train the synaptic weights of the network, minimizing the cost function by adapting weights. The cost function calculates the difference between the desired output and the network output.

2.3.3 NeuroEvolution

When trying to solve a particular task using ANNs, its architecture and the way the model is trained are important decisions to be made. NE tries to solve these problems using GAs to evolve the topology of ANNs [10]. Even though this solution offers several advantages over the backpropagation model, like achieving reasonable solutions in shorter time and being more flexible to factors that are not easily incorporated in the backpropagation method [25], it has two main drawbacks: the competing conventions problem and the topological innovation problem.

2.3.3.1 Competing Conventions Problem

A well-known problem of NE is the Competing Conventions Problem [27], in which two neural networks that order their hidden nodes differently might still be functionally equivalent. This represents a problem because these different representations can produce a flawed offspring during crossover. For example, crossing [A, B, C] with [C, B, A] using single-point crossover, in which one random point is chosen to split each genome and one of the parts is switched between both of them, will produce [C, B, C] and [A, B, A], both losing one third of important information to the final solution. Therefore these damaged solutions will be evaluated by the algorithm needlessly, increasing the computation time inherent to this process.

2.3.3.2 Topological Innovation Problem

Another problem must be taken into consideration if we need to adjust the existing structure, e.g. adding a new connection to the ANN. Such change will decrease fitness before the connection weight has time to optimize itself and reproduce in order for its children to develop a subsequent structural development that can be promising.

2.3.4 NeuroEvolution of Augmenting Topologies

The NEAT algorithm, developed by Ken Stanley, seeks to solve some of the drawbacks of previous NeuroEvolution approaches. NEAT proves to be effective in preserving structural innovation using historical markers, introducing speciation and by applying the principle of incremental growth from minimal structure, through a process called complexification [29].

During crossover, each genome is aligned based on its historical markers: matching genes if they share an ancestral origin, disjoint or excess genes if they exist, respectively, inside or outside the innovation numbers (unique numbers that are incremented for each new gene created) of their parents. Thus, two different parents will manage to find their structure similarities without expensive topological analysis and the extra genes are always inherited from the fittest parent. This process is focused on fighting the competing conventions problem as the functional parameters of each individual are preserved and offspring are undamaged.

Historical markers play another important role in speciation. Topological innovation can be protected this way, by ensuring genomes only compete within a population niche, protecting innovative structures. Two organisms are said to be in the same species if they represent topologically similar networks. The topological disparity between two neural networks can be quantified through the number of excess and disjoint nodes of both genomes and a compatibility function can be employed with a threshold to decide whether two fit individuals are similar enough. If not a new species can be created to give time for both topologies to evolve and compete with similar ones, avoiding being early discarded.

Stanley also defended that starting with random topologies does not lead to finding minimal solutions, since the population already starts with unnecessary nodes, thus starting with minimal structure allows the algorithm to search for the solution in a low-dimensional search space, which improves performance (a process called complexification).

In finance, NEAT has already been applied to stock prediction [26] and, even though the results were not entirely conclusive, in three specific stocks (Apple, Microsoft and Yahoo) it outperformed backpropagation strategy.

2.3.5   Speciation

A GA seeks to find an optimal solution based on an optimization function, after multiple iterations over a procedure of selection, crossover and mutation, as explained before in this report. However this approach brings along a certain loss of genetic diversity, as inevitably the population will converge towards one fit individual and its close mutants, leading to that loss of genetic diversity. In nature, an ecosystem is composed by different physical spaces (niches) exhibiting different features, which allows both the formation and maintenance of different types of life that compete to survive (species). For each niche, the physical resources are finite and must be shared among the individuals belonging to it. This promotes genetic diversity by encouraging the emergence of species in different sub-spaces of the environment (niches) and we experience a wide variety of species co-existing within an ecosystem, evolving at the same time and competing using the environmental resources [8].

The main goal of carrying this concept to the Evolutionary Algorithms relies on the fact that, given some value calculated over the parameters of a particular genome, a threshold could be set to consider that two not looking alike individuals in the population must not be crossing over each other. This way, each individual will compete against individuals of the same species and each will have time evolve their fittest individuals for an optimal solution that took into account genetic diversity. A set of existing speciation methods is presented next.

2.3.5.1 NEAT and Historical Markers

Introduced in [29] and as explained before in this report, historical markers identify the original ancestor of each gene. When comparing a pair of genes, the disjoint genes (D), excess genes (E) and the average weight difference of matching genes ($\overline{W}$) are taken into account to calculate a compatibility distance between the different structures, which is a simple linear combination of them:

$$\delta = \frac{c_1 \times E}{N} + \frac{c_2 \times D}{N} + c_3 \times \overline{W} \qquad (2.1)$$

where N is the number of genes in the larger genome of the population and the coefficients c_1, c_2 and c_3 allow to adjust the relative importance of the three factors (E, D and $\overline{W}$). By setting a compatibility threshold, the result of applying this measure will decide in which existing species it fits, or if a new one must be created.

2.3.5.2 Taxon-Exemplar Scheme

Brooker [3] used taxon and exemplars in his learning algorithm to create a restricted mating policy.

A taxon is a string constructed with the alphabet 1, 0, #, where # matches both a 0 or a 1, and each individual is initialized in the population with it. In his policy, he wanted to restrict mating among similar taxon strings, which are identified by a score calculated between them and a given exemplar binary string. The score is calculated based on the correct matches and the number of #'s in the taxon.

If a certain number of matching taxon strings (strings with a score matching a threshold score) are available within the population, parent strings are randomly chosen from this group, otherwise parent strings are chosen according to a probability distribution, calculated based on the matching score of taxon strings.

2.3.5.3 The Tag-Template Method

Similar to the previous method, a template and tag strings are introduced [9]. The template string is constructed the same way as the taxon, but the tag string is binary, having both the same size. They are created at random in the initial population, along with the functional string, and have the same purpose as the taxon strings: before crossing a pair of individuals, tag and template strings are compared, a matching score is calculated and their mating depends on that score. The difference between this method and the one before stands for its purpose, as tag and template strings corresponding to the fittest individuals in early populations are highlighted and, later on, crossover is only performed between matching strings of each peak.

2.3.5.4   Phenotypic and Genotypic Mating Restriction

Deb [9] has also developed two mating restriction schemes based on the phenotypic and genotypic distance, which are respectively calculated through Euclidean distance and Hamming distance. To choose a mate for an individual in the population, one of these distances is determined and if it is close to a given parameter they will engage in crossover, otherwise another individual is chosen and the distance is computed again. This process goes on until a fitting mate is found and, if this does not happen, the choice is made randomly. These mating restrictions were developed to be applied in multi-modal test problems.

2.3.5.5   Fitness Sharing

One of the most widely used niching methods that treats fitness as a resource to force all individuals within a niche to share a similar fitness between them. This is accomplished by using an adjusted fitness, called fitness sharing, which reduces the fitness of each individual by an amount approximately equal to the number of similar individuals in the population, using a defined set of parameters to measure the similarity between individuals. Formally, the shared fitness $\varphi_{sh,t}(\tau)$ of an individual τ is computed according to the following formula:

$$\varphi_{sh,t}(\tau) = \frac{\varphi(\tau)}{m_t(\tau)} \tag{2.2}$$

where $\varphi(\tau)$ is the raw fitness and $m(\tau)$ is an estimate of the number of individuals belonging to the same niche.

However, this method comes with the drawback that setting proper values for these parameters implicates the need of some deep theoretical knowledge about the fitness landscape [15].

2.3.5.6   Dynamic Niche Sharing

First method to introduce a mechanism to dynamically classify species among the population during the evolution process [21]. At each generation, an algorithm called Dynamic Peak Identification pinpoints the existing niches in the population and classifies new individuals as belonging to one of those, or a special division called nonspecies. Then two fitness sharing methods are applied: one to update the fitness sharing value of the individuals in each niche and the standard one to be applied to the individuals of the nonspecies class. The main drawback of this approach relies on the fact that the number of peaks in the fitness landscape must be previously known.

2.3.5.7 Dynamic Niche Clustering

A method introduced by [13] that uses the concept of nicheset. Each niche is defined by some parameters that change during the evolution process and the method allows for an individual to be a member of more than one niche, which means that they may even overlap in the search space, although some parameters are delimited to avoid the excessive growth or decrease of the number of niches along the process. This overlapping of niches is a controlled process called striation that does not allow niches to converge by penalizing members belonging to both niches, even though an improved version of this method introduces a function to decide whether niches must be merged or split during evolutionary process [14]. Overall, the method relies on the assumption that the search space must have bounds of the same magnitude, otherwise niche radius can be wrongly calculated, and the initial population stands as an important aspect to define the bounds of the niche radius. The flexibility of not constraining individuals to the apex of the niche makes natural that their fitness is lightly reduced compared to other techniques.

2.3.5.8 Adaptive Species Discovery

A niching method that removes the premise of perfect discrimination which serves as base for niching techniques, like Fitness Sharing, and is more effective in handling with irregular spaced peaks in a multi-modal landscape. Like Dynamic Niche Sharing, this method does not require any a priori knowledge, but complements it as information about the fitness landscape is not needed as well. Each niche is identified by a representative individual, which is chosen among the members, and the explicit knowledge about the number of niches and their positions in the search space grants a way to preserve species masters between generations, in an elitist process that maintains the number of niches. The discovery process is based upon a set of information dynamically acquired at each generation, through a Hill-Valley function [30], which allows the detection of genetic divergence within the population. The function was first adopted to verify whether two points lie or not in the same peak in the landscape, however it also grants the detection of genetic divergence phenomena within an artificial population, being the main idea of adopting this function to prevent individuals with very different genotypic features from being considered members of the same species [8].

2.4 Pattern Detection

One of the most difficult tasks for investors is to analyze financial charts to detect patterns. Three main concepts must be distinguished regarding this subject: time series, pattern recognition and pattern discovery.

A time series stands for a collection of observations chronologically made [11], while in pattern recognition the goal is to identify known patterns on the time series and pattern discovery consists in finding new patterns that occur in the time series and are unknown [5].

The appearance of chart patterns in financial time series is often a premeditation of a possible change in price trend. For the purpose of identifying these patterns, various approaches can be used to measure the similarities between chart pattern templates and sequences extracted from financial time series. Despite of the method chosen for pattern matching, there is not an ambiguous standard for how these patterns should be defined. Some well-known patterns are frequently used to evaluate the effectiveness pattern matching methods, but, once again, as there are no precise definitions for these chart patterns, researchers and financial analysts may specify them in the most suitable way for their applications.

2.4.1   Chart Patterns

In a financial time series, some similar graphic representations appear over time and, as these graphics are representative of behavior patterns of investors, as well as the market performance is influenced by them, some of these similarities are good warning signs on what the market will do next. Patterns can be characterized as variable or fixed fluctuation patterns, whenever they have an uncertain or fixed number of highs and lows respectively. They can also be identified over the continuous market signal line or in the candlestick graphic representation and most of them can be derived into two patterns of similar behavior, except for the fact that one represent an uptrend or ascending movement and the other a downtrend or descending movement, i.e. tops and bottoms, which are formations that represent a price level of resistance and support, indicating a turnabout in the trend.

Patterns can be found within the continuous signal of the stock market, which indicates the closing price in each time interval, or within the candlestick representation signal, which, as shown in Fig. 2.1, also indicates the opening price, highest value and lowest value, besides the closing price.

Taking this into account, some the most known, used and reliable chart patterns are presented next [4, 31].

2.4.1.1   Ascending and Descending Triangle

This pattern is characterized by having an horizontal top and an up-sloping bottom trend lines, if it is the ascending triangle, or a down-sloping top and horizontal bottom trend lines, in case of being a descending triangle. Ascending triangles usually confirm an uptrend and are more reliable when found in bullish market conditions, while descending triangles appear to support a downtrend and are more reliable in

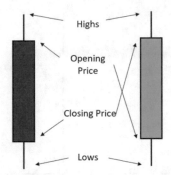

Fig. 2.1 Candlestick representation

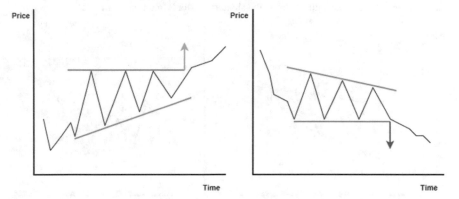

Fig. 2.2 Graphic representation of ascending and descending triangles charts

bearish market conditions. Figure 2.2 shows the graphic representation of *Ascending and Descending Triangles*.

2.4.1.2 Head and Shoulders Tops and Bottoms

This pattern can have multiple heads or multiple shoulders (rarely both) when it represents bottoms, being the head(s) always lower than the shoulders. In the case of being a head and shoulders tops, it has multiple shoulders and they must mirror themselves related to the head, being this last one always higher than the shoulders. It is a reversal type and comes along with an indication of change in trend. Figure 2.3 shows the graphic representation of *Head and Shoulders Tops and Bottoms*.

2.4.1.3 Rising and Falling Wedges

This pattern is characterized by two trend lines that must both slope upward or downward, depending if we are considering rising or falling wedges respectively,

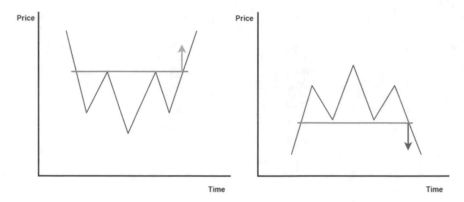

Fig. 2.3 Graphic representation of head and shoulder bottoms and tops charts

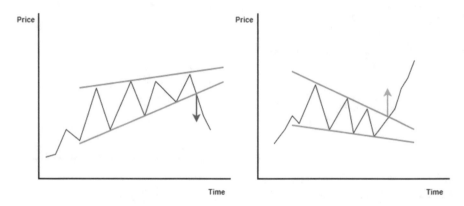

Fig. 2.4 Graphic representation of rising and falling wedges charts

where in the first case the bottom trend line must be steeper than the top trend line and otherwise in the second case. The minimum duration of these patterns is 3 weeks and usually don't last more than 3 months. Rising wedges are associated with bearish market while falling wedges tend to appear more in a bullish market. Figure 2.4 shows the graphic representation of *Rising and Falling Wedges*.

2.4.1.4   Flags and Pennants

Both patterns are followed by ascent or descent burst, whether they represent a confirmation of an uptrend or downtrend, respectively. In the case of flags depending on whether we are considering a bullish or bearish market, the pattern oscillates between two horizontal lines with an upward or downward slope. The flags are often called Bull Flag or Bear Flag depending on the direction of the tendency. On the other hand, the pattern in the pennants oscillates between two intersecting trend

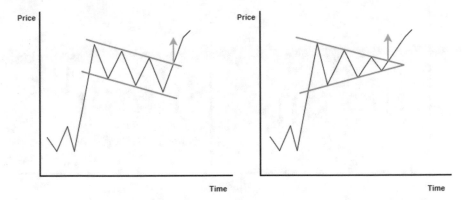

Fig. 2.5   Graphic representation of flag and pennant charts

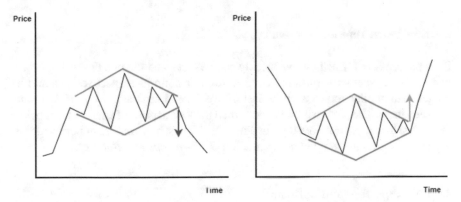

Fig. 2.6   Graphic representation of diamond tops and bottoms charts

lines, one sloping downward from the top and the other sloping upward from the bottom. Figure 2.5 shows the graphic representation of a *Flag* and a *Pennant*.

2.4.1.5   Diamond Tops and Bottoms

Both patterns are characterized by having higher top points and lower floor points in the first part of the pattern, followed by lower top points and higher floor pints in the second part of the pattern. The difference between them relies on the start price trend. A diamond tops pattern has its start price in an upward trend and the diamond bottoms pattern is characterized by a starting price in a downward trend. Figure 2.6 shows the graphic representation of *Diamond Tops and Bottoms*.

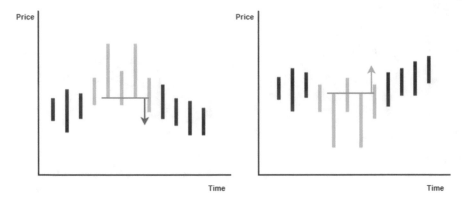

Fig. 2.7   Graphic representation of horn tops and bottoms charts

2.4.1.6   Horn Tops and Bottoms

Both patterns are formed by five weekly candlesticks and the two upward or downward spikes, whether we are referring to tops or bottoms respectively, should be
longer than similar spikes over the last 48 weeks and separated by a smaller stick
between them. The pattern is confirmed when the lowest price of stick 5 closes below
the lowest low, in case of horn tops, or above the highest high, in case of horn bottoms.
Figure 2.7 shows the graphic representation of *Horn Tops and Bottoms*.

2.4.1.7   Pipe Tops and Bottoms

Both patterns are formed by four weekly candlesticks and have two adjacent upward
or downward price spikes, whether we are referring to tops or bottoms respectively.
They both should be longer than similar spikes over the last 48 weeks and the pattern
is confirmed when the price closes below their lowest low, in case of pipe tops,
or above their highest high, in case of pipe bottoms. Figure 2.8 shows the graphic
representation of *Pipe Tops and Bottoms*.

2.4.2   Detection of Perceptually Important Points

One of time series pattern matching techniques is PIPs [12], which allows a big
reduction in the dimension of time series, maintaining the main characteristics of
data. These points are important for the identification of patterns as they stand for
the ones that characterize the time series. Pattern recognition after defining the PIPs
can be done using two techniques: template-based or rule-based. The template-based
approach allows a point-to-point comparison between time series and patterns, as the

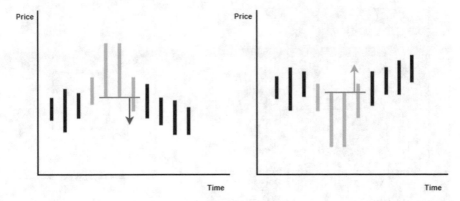

Fig. 2.8 Graphic representation of horn tops and bottoms charts

rule-based approach stands for the explicit relation between time series and patterns [20].

2.4.3 Application of Templates

The most straightforward way to achieve a comparison measure between two patterns is through template matching. By defining the shape of query patterns (pattern templates) visually, point-to-point direct comparisons can be performed [11]. However, different segments of a time series can have different amplitudes so, for instance, after identifying the PIPs a normalization is necessary in order to simplify the identification of identical patterns. Then, a direct point-to-point comparison can be done between the input sequence of points and the query sequence representative of the template.

2.4.3.1 Grid of Weights

One other way to identify patterns through the application of templates is using a grid of weights [19], which stands for a 10x10 grid, like the one in Fig. 2.9, which allows the recognition of patterns by assigning higher classification to the cells of the grid matching the template, and by defining a minimum fit value threshold for which the sum of scores of the cells matching the close price signal must achieve.

This has the disadvantage of being a subjective process in which the threshold value must be adjusted, taking into consideration that higher values can miss price windows that fit the pattern and lower values can lead the to the identification of false patterns. However, this subjectivity combined with the fact that the weights configuration can be changed, may grant the possibility of going further after the

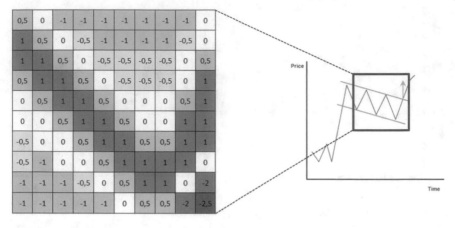

Fig. 2.9 Bull flag template from Leigh, Paz, and Russell [19]

needs that each specific template require to be identified with more precision, even allowing to take into account not only closing prices but also highs, lows and opening prices.

In finance, some works have been developed relying on this representation for pattern detection. Cervelló-Royo [6] and Arévalo et al. [1] took what was said before into consideration and developed a grid with weights configured to support a scheme where the body of candlesticks were used to detect the Bull Flag pattern, as shown in Fig. 2.10. The price should always visit the cell labeled 5 and if the threshold was set to 3, the body of the candlesticks should only visit negative cells twice. Parracho [24] used only the closing price, but tested this representation with 4 different patterns (uptrend, downtrend, breakout and bull flag) combined be-tween them. Both approaches showed results that outran the market and Parracho's version of the grid also outperformed Buy and Hold.

2.4.4   Application of Rules

Through this approach, rules are defined to characterize the shape of wanted patterns. One of the disadvantages of template based approach is, as stated before, that the relationship between points is hard to define unequivocally and sometimes imprecise, so this methodology claims to overcome that problem specifying the relative positions of the points composing the pattern, e.g., for a Head and Shoulders pattern, both shoulders must be at a similar height and at a lower level than the head.

It is not possible to guarantee the reliability of these specifications with a template based approach because similar shaped patterns may be found that violate these rules. This way, given a certain number of points (e.g. PIPs) the matching pattern is depending on the validation of all rules.

Fig. 2.10 Bull flag template from Cervelló-Royo, Guijarro, and Michniuk [6]

0	0	0	0	0	0	0	0	0	0
0	0	0	0	0	0	0	0	0	0
0	0	0	0	0	0	0	0	0	0
0	0	0	0	-1	-1	-1	-1	-1	-1
0	0	0	-1	-2	-2	-2	-2	-2	-2
0	0	-1	-3	-3	-3	-3	-3	-3	-3
0	-1	-3	-5	-5	-5	-5	-5	-5	-5
0	-1	-5	-5	-5	-5	-5	-5	-5	-5
0	-1	-5	-5	-5	-5	-5	-5	-5	-5
5	-1	-5	-5	-5	-5	-5	-5	-5	-5

Despite this technique has overall poor matching results for the majority of patterns, some specific ones like Head and Shoulders and Triple Top can be detected with distinction [11].

2.5 Chapter Conclusion

In Table 2.1 are presented the results of some relevant studies for the developed work, which apply some of the previous presented techniques to the stock market. Notice that presented results are not sorted in any particular order as it is difficult to rank them while being applied over different markets and testing periods. Although these comparisons are difficult to make, for each particular case, either using GA [5, 24], speciation [26], grid template detection approach [6, 19, 24] or other template detection approaches [11, 20], the results showed improvements against competitor methodologies. Based on these results, the main intent of this work is to develop an algorithm that aggregates and takes advantage of some aspects of all these approaches: combination of GA, speciation and pattern detection using grid template approach.

Table 2.1 Results of some relevant studies for this report

Ref and year	Method	Financial market	Data	Period	Algorithm performance	Comparator performance
[26] 2015	NEAT	Apple/Microsoft/ Yahoo	Stock price	2012–2014	56.81%/13.19%/ 16.69% (return on investment: best result)	22.10%/0.50%/ 16.27% (backpropagation method)
[5] 2013	SAX + GA	S&P 500	Stock price	1998–2010	37,60% (return on investment)	0,32% (buy and hold: return on investment)
[20] 2016	PIPs + SAX	S&P 500	Stock price	2010–2014	71,77% (total avg. return)	61,22% (buy and hold: total avg. return)
[19] 2002	Bull flag pattern w/grid template	NYSE composite index	Stock price	6/08/1980– 15/09/1999	6,24% (trading rule avg. return)	3,91% (market avg. return)
[6] 2015	Bull flag pattern w/grid template	DJIA (Dow Jones Industrial Average)	Candlesticks	22/5/2000– 29/10/2013	37,53% (total return)	NA
[11] 2007	Template based versus rule based	HSI (Hong Kong Hang Seng Index)	Stock price	2532 data points	98%/48% (template/rule: avg. accuracy)	96% (PAA: avg. accuracy)
[24] 2011	Uptrend and downtrend pattern w/grid template + GA	S&P 500/DJIA/NYSE composite index	Stock price	1998–2010	36.92%/16.33%/ 10.02% (profitability)	−4.69%/0.12%/ −1.41% (buy and hold: profitability)

References

1. Arévalo, R., García, J., Guijarro, F., Peris, A.: A dynamic trading rule based on filtered flag pattern recognition for stock market price forecasting. Expert Syst. Appl. **81**, 177–192 (2017)
2. Bäch, T.: Evolutionary Algorithms in Theory and Practice. Oxford University Press, New York (1996)
3. Brooker, L.B.: Doctoral Dissertation: Intelligent behavior as an adaptation to the task environment, University of Michigan (1982)
4. Bulkowski, T.N.: Encyclopedia of Chart Patterns, 2nd edn. Wiley, New Jersey (2011)
5. Canelas, A., Neves, R., Horta, N.: A SAX-GA approach to evolve investment strategies on financial markets based on pattern discovery techniques. Expert Syst. Appl. **40**, 1579–1590 (2013)
6. Cervelló-Royo, R., Guijarro, F., Michniuk, K.: Stock market trading rule based on pattern recognition and technical analysis: forecasting the DJIA index with intraday data. Expert Syst. Appl. **42**, 5963–5975 (2015)
7. Chang, T., Yang, S., Chang, K.: Portfolio optimization problems in different risk measures using genetic algorithm. Expert Syst. Appl. **36**, 10529–10537 (2009)
8. Cioppa, A.D., Marcelli, A., Napoli, P.: Speciation in evolutionary algorithms: adaptive species discovery, GECCO '11. In: Proceedings of the 13th Annual Conference on Genetic and Evolutionary Computation, pp. 1053–1060 (2011)
9. Deb, K.: Genetic algorithms in multimodal function optimization. Master's Degree Thesis, University of Alabama, USA (1989)
10. Floreano, D., Durr, P., Mattiussi, C.: Neuroevolution: from architectures to learning. Evol. Intell. **1**(1), 47–62 (2008)
11. Fu, T., Chung, F., Luk, R., Ng, C.: Stock time series pattern matching: template-based vs rule-based approaches. Eng. Appl. Artif. Intell. **20**, 347–364 (2007)
12. Fu, T., Chung, F., Luk, R., Ng, C.: Representing financial time series based on data point importance. Eng. Appl. Artif. Intell. **21**, 277–300 (2008)
13. Gan, J., Warwick, K.: A genetic algorithm with dynamic niche clustering for multimodal function optimization. In: Proceedings of IEEE Congress on Evolutionary Computation, pp. 248–255 (1998)
14. Gan, J., Warwick, K.: Dynamic niche clustering: a fuzzy variable radius niching technique for multimodal optimisation in GAs. In: Proceedings of IEEE Congress on Evolutionary Computation, pp. 215–222 (2001)
15. Goldberg, D.E., Richardson, J.: Genetic algorithms with sharing for multimodal function optimization. In: Proceedings of the Second International Conference on Genetic Algorithms on Genetic Algorithms and Their Application, pp. 41–49 (1987)
16. Gorgulho, A., Neves, R., Horta, N.: Applying a GA kernel on optimizing technical analysis rules for stock pricing and portfolio composition. Expert Syst. Appl. **38**, 14072–14085 (2011)
17. Holland, J.H.: Adaptation in Natural and Artificial Systems: An Introductory Analysis with Applications. The MIT Press, Cambridge (1992)
18. Kearney, W.T.: Using genetic algorithms to evolve artificial neural networks. Honors Thesis in Computer Science, Colby College, USA (2016)
19. Leigh, W., Paz, N., Russell, P.: Market timing: a test of a charting heuristic. Econ. Lett. **77**, 55–63 (2002)
20. Leitão, J., Neves, R., Horta, N.: Combining rules between PIPs and SAX to identify patterns in finantial markets. Expert Syst. Appl. 242–254 (2016)
21. Miller, B., Shaw, M.J.: Genetic algorithms with dynamic niche sharing for multimodal function optimization. In: IEEE International Conference on Evolutionary Computation, pp. 786–791 (1996)
22. Mitchell, M.: An Introduction to Genetic Algorithms. The MIT Press, London (1996)
23. Murphy, J.J.: Technical Analysis of the Financial Markets: A Comprehensive Guide to Trading Methods and Applications, New York Institute of Finance (1999)

24. Parracho, P., Neves, R., Horta, N.: Trading with optimized uptrend and downtrend pattern templates using a genetic algorithm kernel. IEEE Congr. Evol. Comput. 1895–1901 (2011)
25. Qiu, M., Song, Y., Akagi, F.: Application of artificial neural network for the prediction of stock market returns: the case of the Japanese stock market. Chaos, Solut. Fractals **85**, 1–7 (2016)
26. Sandstrom, C., Herman, P., Ekeberg, O.: An evolutionary approach to time series forecasting with artificial neural networks. Bacherlor's Thesis at KTH Royal Institute of Technology, Stockholm, Sweden (2015)
27. Schaffer, J.D., Whitley, D., Eshelman, L.J.: Combinations of genetic algorithms and neural networks: a survey of the state of the art, COGANN-92. In: International Workshop on Combinations of Genetic Algorithms and Neural Networks, pp. 1–37 (1992)
28. Silva, A., Neves, R., Horta, N.: A hybrid approach to portfolio composition based on fundamental and technical indicators. Expert Syst. Appl. **42**, 2036–2048 (2014)
29. Stanley, K.O., Miikkulainen, R.: Evolving neural networks through augmenting topologies. Evol. Comput. **10**(2), 99–127 (2002)
30. Ursem, R.K.: Multinational evolutionary algorithms. In: Proceedings of the Congress on Evolutionary Computation, pp. 1633–1640 (1999)
31. Wan, Y., Si, Y.: A formal approach to chart patterns classification in financial time series. Inf. Sci. **411**, 151–175 (2017)
32. Werbos, P.J.: Beyond regression: new tools for prediction and analysis in the behavioral sciences. Ph.D. Thesis, Harvard University (1974)
33. Wilder, J.W.: New concepts in technical trading systems. Trend Research: Indiana, USA (1978)

Chapter 3
Architecture

3.1 System's Architecture

To determine optimal timings for entering or leaving a specific market based on the analysis of template chart patterns optimized using a GA with speciation, while taking also into consideration the risk associated with the capital invested, Fig. 3.1 presents the overall overview of the system's architecture designed for that purpose.

It follows the most common logical multi-layered architecture composed by three specific layers: the presentation layer, the application layer and the data layer. This gives the ability to develop and update the technology stack of one layer without impacting the remaining ones, providing flexibility to scale the application, as well case of maintenance regarding the code base. The execution of the above system can be interpreted through the following steps:

1. User specifies the input parameters to be implanted in the system and used by the lower levels, such as financial data, timing constraints and Trading Algorithm parameters;
2. A component responsible for Pre-Processing the information provided by the user gathers all the needed data from the Financial Data component, normalizes it into the standard representations of the system and populates the adequate data structures to provide all the tools for proper operation of the Trading Algorithm in the next stage;
3. The Trading Algorithm component will be continuously generating a trading signal relying on a template-based pattern detection technique that is trained with GA with speciation and also outputting the most relevant results during its execution;
4. At the end of the system's execution, the results of the Trading Algorithm component will be presented to the user for analysis.

All these architectural components will be addressed with more detail in the following subsections.

© The Author(s), under exclusive license to Springer Nature Switzerland AG 2021   25
T. Martins and R. Neves, *Stock Exchange Trading Using Grid Pattern Optimized by A Genetic Algorithm with Speciation*, SpringerBriefs in Computational Intelligence,
https://doi.org/10.1007/978-3-030-76680-1_3

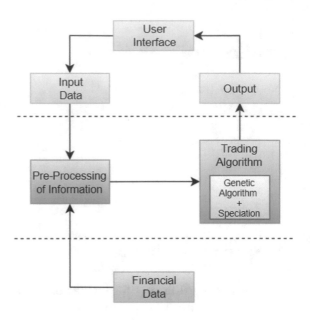

Fig. 3.1 System's architecture by components

3.2 User Interface

The interaction with the program provided for the user is mainly performed via command prompt, as the main goal of this project relies on proving a working trading algorithm and not a user friendly interface rigged with all its heuristics. The user can use it not only to issue inputs to setup and run the algorithm, as well as to choose between various ways of presenting results.

To run the algorithm a set of input parameters are displayed and editable in any way each user finds suitable depending on the analysis intended to be done. These inputs are written in a configuration file that will be read by the program upon starting, and are supported by a manual on prompt, helping users in understanding the meaning of each parameter. These parameters are mostly GA related, as the way some genetic operators act depend on these values, however one in particular is an exception and must be highlighted for the reason that its existence is merely for testing purposes and is related to the program capability of operating in 3 different ways: (1) to simulate the proposed system; (2) to simulate a system without speciation; (3) to simulate the traditional system proposed by [2].

To analyze results, at least one run of the program must have been executed prior to analysis and 3 options are available:

- Export the results into a *.xlsx* file where relevant information regarding each stock is provided and can be easily manipulated by programs such as Microsoft Excel;
- Display graphs containing the training and testing phases of the algorithm, for a specific run and a specific stock. These graphs contain the candlestick representation of every day in the periods related to each stage of the algorithm, along with

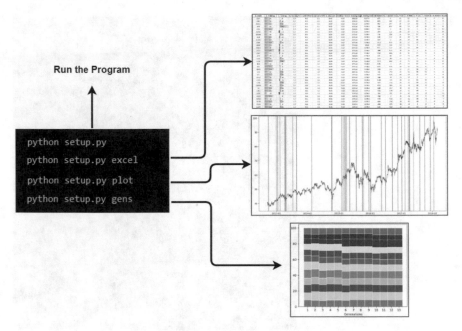

Fig. 3.2 Commands issued to run the program or present results

indicators for market entries and exits, whether they are positive or negative. The frameworks *matplotlib* and *mpl_finance* were used for this purpose.

- If a run of the algorithm using speciation was made, an horizontal graph bar can be displayed to visually illustrate the evolution of the fittest individual of each specie throughout all the generations. *Matplotlib* also supports this feature of the program.

Figure 3.2 presents a figurative example of running commands to execute the program or analyze results.

3.3 Financial Data

The program obtains and stores data from companies that is going to be used by the Trading Algorithm component. The process of collecting data uses a python module called *alphavantage*, a provider of APIs for real time and historical data on stocks. The set of companies used in this research belong to Standard and Poor's 500 (S&P 500) and the API allows the download of daily data from the companies since the year 2000 up to the current day. Each data entry consists of a set of opening and closing prices, highs and lows, and volumes of transactions during each considered period.

Fig. 3.3 CSV file from
NFLX stock

date	1. open	2. high	3. low	4. close	5. volume
2002-05-23	16.19	17.4	16.04	16.75	7485000.0
2002-05-24	17.0	17.15	16.76	16.94	793200.0
2002-05-28	16.99	17.25	16.2	16.2	472100.0
2002-05-29	16.3	16.3	15.2	15.45	482700.0
2002-05-30	15.51	15.51	15.0	15.0	725300.0
2002-05-31	15.1	15.1	15.0	15.07	604600.0
2002-06-03	15.12	16.09	15.07	15.8	225100.0
2002-06-04	15.9	15.96	15.55	15.65	221800.0
2002-06-05	15.55	16.23	15.5	16.06	109400.0
2002-06-06	16.1	17.25	16.08	16.55	164700.0
2002-06-07	16.49	16.49	15.45	15.66	97800.0
2002-06-10	15.89	16.45	15.89	16.19	34600.0
2002-06-11	16.19	16.64	15.8	16.15	71700.0
2002-06-12	16.15	16.55	15.25	15.3	128500.0
2002-06-13	15.46	15.71	15.12	15.16	183400.0
2002-06-14	15.15	15.2	13.72	13.81	341700.0
2002-06-17	14.19	14.83	12.85	12.91	346800.0
2002-06-18	13.14	13.15	11.79	12.75	772000.0
2002-06-19	12.76	13.95	12.5	13.32	391600.0
2002-06-20	13.6	14.56	13.3	13.4	436500.0
2002-06-21	13.75	13.85	13.27	13.71	167600.0
2002-06-24	13.71	14.2	13.4	13.7	183400.0
2002-06-25	13.84	14.1	13.26	13.5	281500.0
2002-06-26	13.49	13.95	12.9	13.84	208900.0

Regarding the storage of data, it is made locally in Comma Separated Values
(CVS) files, as the one shown in Fig. 3.3 regarding Netflix (NFLX) stock, which
are exported and imported to the program using another Python module named
pandas, widely used for manipulating complex data collections within Python.
Some mechanisms are implemented to extract specific subsets of data, which gives
flexibility for the program to explore various scenarios using different time intervals
of data. Calls to the API are only made if the desired data is not yet in local storage.

3.4   Data Flow and Data Processing

In terms of data being introduced, produced and extracted from the program, we can
divide it in 3 types:

1. Input Data, that contains program variables with predefined values by the user
   via configuration file;
2. Main Program, containing training and testing phases. Training phase is where
   the GA finds the best individual to be tested in Testing phase, and both together
   rely on all the input data to generate relevant information that will be withdrawn
   from the program as output data;

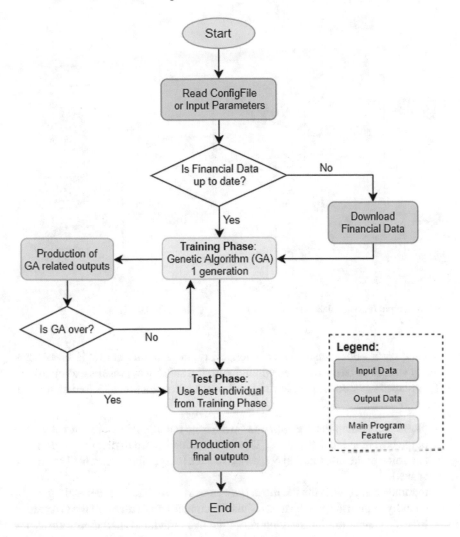

Fig. 3.4   Flow of data chart during program execution

3. Output Data, composed by every piece of information that is extracted from the program, either during the Training Phase of the program (to evaluate the outputs of the GA) or after the Testing Phase to reveal the final results.

Figure 3.4 illustrates the entire process described before.

Regarding the production of outputs, for an easier analysis of the results all retrieved data is organized in a folder hierarchy as follows:

• First layer—Composed by a folder per stock analyzed by the program;

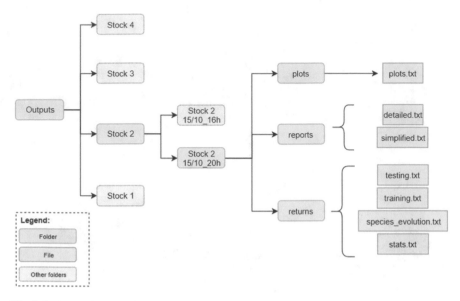

Fig. 3.5 Output files organization

- Second layer—For each analyzed stock, every program output of it is saved separately with dates on each folder, which facilitates subsequent results examinations;
- Third layer—For each specific run involving a particular stock, 3 folders are created:
  - plots—containing the file *plots.txt* where the records related to market entries of the best individual are saved. This information is used to plot output graphics and some of the information contained in excel output file is extracted from here as well.
  - reports—can contain the file *detailed.txt* or *simplified.txt*, depending on the quantity of information the it has. This file contains a full report of the GA stating for each generation information about the live individuals and their returns, the species, the elitists and current best individual, as well as a final more specific insight of the fittest individual returned by the GA. The difference between both files relies on the fact that *detailed.txt* shows the state of the population between each stage of the generation (selection, crossover, mutation, elitism) and *simplified.txt* does not.
  - returns—contains 3 files (*testing.txt*, *training.txt* and *stats.txt*) with information only used by the excel output file, and also *species_evolution.txt* containing information about the best individuals of each species in each generation used by the graph that shows the evolution of the species along the algorithm.

Figure 3.5 exhibits the organization of the output files.

3.5   Trading Algorithm

This stands for the critical component of the system as it will contain the most needful mechanisms to detect chart patterns, i.e. the weighted grid alongside the GA, to train/optimize the detection procedure and support the decision of buying or selling positions in the market.

3.5.1   Pattern Detection

The pattern detection technique follows a heuristic based on templates, namely the 10x10 grid of weights approached in Sect. 2 [4]. Four conditions are being set for each column of the grid for a Bull Flag pattern:

1. the sum of the scores contained in each column must be 0;
2. there must be two cells with a score equal to 1, which is the maximum score allowed;
3. these two cells must serve as an axis of symmetry between the values of the remaining cells;
4. for each cell that is further away from that axis, the score must be lower or equal than the one that is closer.

As an example for this last condition, if in a certain column the third and fourth cells contain the score 1, the second and fifth cells must contain the same value, smaller than 1; first and sixth cells must also contain the same value, smaller than 1 and smaller or equal to the values contained in the second and fifth cells; and for the cells between seventh and tenth positions, the values of each cell must be lower than 1, lower or equal than all first 6 cells, and smaller between them as they are further away from the closest cell belonging to the axis of symmetry i.e. cell 4.

To detect patterns in the traditional way using a grid of weights, first it must be fitted into the price window we want to check and the closing price of each entry will coincide with one specific cell of the grid. The scores of each cell are assigned so the closer the signal is from the pattern, the higher its score is, and the goal of this process is to obtain a fit value (threshold) to show the level of matching between the pattern matrix and the price window. The choice of the threshold value must be done taking into account that low thresholds may lead to false positives and high ones may discard price windows that would fit the pattern, so a middle term must be found which is not always an easy task.

The pattern being detected is the Bull Flag pattern as shown in Fig. 2.5. When trying to detect it using the traditional grid of weights, in addition to the issue of choosing the most suitable threshold, the fact that the grid adapts itself to the price window so each point of the signal is within it, i.e. each closing price of the price window being analyzed, is kept within the limits of the grid, may also lead to find false positives. A flag is only a continuation pattern if it forms an angle that is clearly

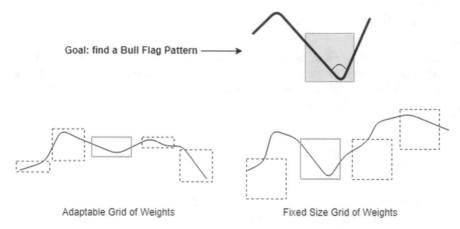

Goal: find a Bull Flag Pattern ──────▶

Adaptable Grid of Weights   Fixed Size Grid of Weights

Fig. 3.6 Difference between an adaptable and a fixed size grid of weights

against the current trend, but as it can be observed in Fig. 3.6 an adaptable grid that varies in size according to the signal may mislead the correct pattern identification, e.g. in cases where the height of each row of the grid is small and the angle formed between the consolidation phase and continuation of the uptrend is very wide.

To overcome both these problems and increase precision while using a grid of weights to detect patterns in the time series, two modifications to the traditional grid of weights are being made, leading to the creation of the Fixed Size Grid of Weights (FSGW):

1. To ensure a choice of a suitable threshold, more information than just the closing price will be used, so that even using a low threshold value the matchings are more reliable. For that purpose, instead of using the grid of weights against the continuous signal of the stock market, which indicates only the closing price in each time interval, the candlestick representation is adopted, which gives information about the highest value, lowest value, opening and closing prices for a specific time interval. The specific way this information is used will be explained in detail next.
2. To avoid the inaccurate detection of flags concerning its shape relatively to the ongoing trend, a fixed size grid is adopted, being the length of each grid row set as an input parameter. As shown in Fig. 3.7, this grid does not depend on the relative size of the signal within the price window, being the positioning of the first column the only imposed constraint and it must be done in a way that fits the grid into the best possible position within the signal to find a Bull Flag, whether the following columns end up being Bull Flag match or not. This way, all pattern matches will be similar even in its relative size in the time series.

Regarding the use of candlestick representation, even though we have more information than the closing price now, not everything is of the same importance for calculating the final score, which is why information must be divided into groups

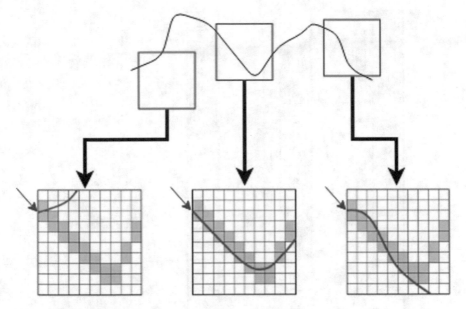

Fig. 3.7 Constraint positioning of the first column of the grid

and each must be weighted differently according its importance in the grid column score for that specific time interval.

Formally, consider a cell named A, corresponding to where the closing price of the signal hits in that grid column; a set of cells named B, corresponding to the cells between where the opening and closing prices hit for a specific column of the grid (excluding the closing price cell); and a set of cells named C, corresponding to the cells visited by the signal during that time interval of the price window and excluding cells contained in A and B. Consider also three weights w_1, w_2 and w_3, which are assigned to each the previous stated groups of cells, so the final score ϕ for that specific column is given as:

$$\phi = A \times w_1 + \sum_0^i \frac{B_i}{size(B)} \times w_2 + \sum_0^i \frac{C_i}{size(C)} \times w_3 \qquad (3.1)$$

being $w_1 > w_2 > w_3$ and $w_1 + w_2 + w_3 = 1$. If B or C are empty sets, their weights (w_2 and/or w_3) are added to w_1 and become 0. The weights w_1, w_2 and w_3 are randomly assigned to each individual by the algorithm during its creation and optimized by the GA from then on, all this as long as the previous rules are met.

Figure 3.8 illustrates the process described above, where for the same column in the grid, three different candlesticks are presented with the same closing price and considering the following weights: $w_1 = 0.5$, $w_2 = 0.3$ and $w_3 = 0.2$.

As it is observable, while taking only closing prices into account the score for each column would be 1 in the traditional way, but if we add to the balance more

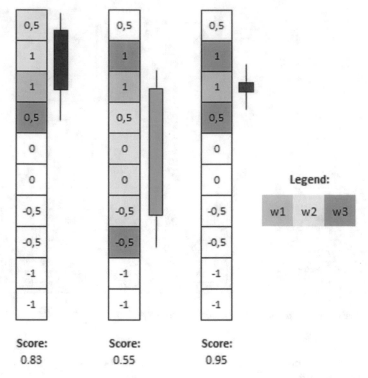

Fig. 3.8 Illustration of weighted scores procedure

information than just the closing price, the risk associated with a momentary increase of market volatility (reflected by a big candlestick) is easily tracked as it results in a lower score.

3.5.2   API for Genetic Algorithm

The most commonly used Python framework for rapid prototyping and testing of ideas based on GA is called DEAP [3]. It has already implemented genetic operators, such as crossover, mutation and selection, and also supports the creation of specific genetic operators (in the scenario of this report a speciation operator would be needed). Besides this, DEAP also offers a wide range of representations for GA's individuals (List, Array, Set, Dictionary, Tree, Numpy Array, etc).

When using a FSGW, some aspects regarding its representation can be already outlined:

- Each individual will necessarily be composed by a list of various sublists;
- Each sublist can vary in size, depending on the genetic encoding;

- Genetic operators such as crossover and mutation will involve the manipulation of entire sublists or their positions, and the ones chosen can be different in each iteration of the algorithm.

With these pointed out and although DEAP's features allow adaptation to different representations, the level of complexity of trying to represent and manipulate a FSGW using this framework would be at least as challenging as creating an API from scratch, which ended up to be the solution found to address this issue.

In the following subsection the genetic encoding used and the way the API manipulates data in each genetic operator will be explained in detail.

3.5.2.1 Design Decisions

To effectively handle and manipulate all used data structures a hybrid approach using procedure-oriented and object-oriented programming paradigms were used. As individuals and species are the two structures that are the most used and manipulated, both represent objects in the program. On the other end, genetic operators are designed around functions that support them, representing the procedure-oriented part of the program.

Each individual contain an unique identifier, its genetic encoding structure, training ROI value, adjusted fitness and the generation in which it was created. Each specie has also an unique identifier, two values that limit which individuals' compatibility values are accepted within this specie, two lists containing the entire species' individuals and the elitists and a constant that holds the number of offsprings, i.e. number of individuals allowed to be carried into the next generation for that specific specie. Some of the attributes stated before are for GA purposes and their usefulness will be explained with more detail in the following subsections.

3.5.3 Genetic Algorithm and Speciation

The main goal of using a GA is to find scores for the grid matrix and values for weights w_1, w_2, and w_3 that maximize the ROI in the training period. The resulting grid from this optimization procedure will then be applied in the testing period. Figure 3.9 presents a schema of the GA information flow using system's genetic encoding and genetic operator's way of manipulating individuals.

3.5.3.1 Genetic Encoding

As the 10x10 grid being used to represent each individual of the population is composed by 100 cells, it would be appropriate to find a way of saving computational resources. Therefore, instead of having a data structure containing all the scores of

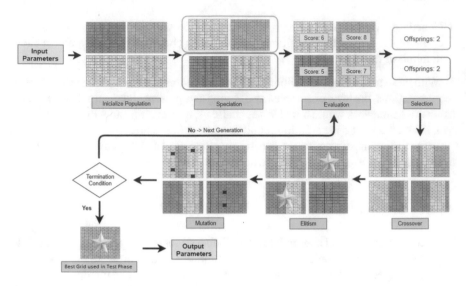

Fig. 3.9 Genetic Algorithm schema

each position of the grid, it is a better approach to take advantage of the symmetry property stated before.

The genetic encoding is then going to use the following representation: for each column of the grid, the first positions will contain two values corresponding to the number of cells above and below the ones that represent the axis of symmetry; the following positions will contain all different scores the respective column contains; the last three positions after all the columns have been handled will contain the weights w_1, w_2, and w_3.

For the API, an individual is represented by a list with 11 positions, being the first 10 composed by elements of each column of the grid and the last a dictionary containing the 3 weights associated to it. Each column is also a list, being its first position a sublist of it. As stated before, it contains two values representing the number of cells above and bellow the axis of symmetry and the highest of this values determine the size of the list for that column. The remaining positions of the column list are filled with each different score that column holds.

Figure 3.10 serves as example to illustrate how, for the case of the Bull Flag pattern detection grid, we can build a genetic encoding representing a whole grid of 100 cells, saving 23 positions in the structure of each individual.

3.5.3.2   Speciation

The goal of using speciation in this work is to maintain diversity in the population and protect new individuals against biased solutions that optimize faster. For this purpose, a compatibility function will be used taking into account the weights that

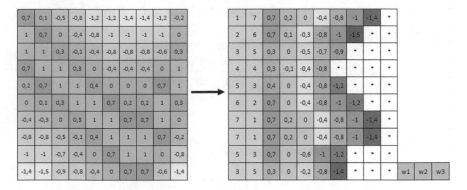

Fig. 3.10 Genetic Encoding of the Grid and Weights: **a** Normal grid representation; **b** Chromosome's genetic encoding representation

have been assigned to each individual. The candidates will then be divided in species, according to a compatibility value calculated using the following equation:

$$\delta = w_1 \times A + w_2 \times B + w_3 \times C \tag{3.2}$$

where w_1, w_2, and w_3 correspond to the weights of each individual and A, B and C are constants set to 0.65, 0.225 and 0.125 respectively. These constants are used to highlight variations in the weights that represent a significant impact in each individual scores, i.e. to w_1 is assigned the highest constant as this weight is the most representative of the three, and so on.

The compatibility values will be comprehended between 0 and 1 and bound values must be set to define each species. For example, if bound values are set to be multiples of 0.025, we can define a species X that will comprehend all the individuals with compatibility values between 0.4 and 0.425, a species Y with compatibility values between 0.425 and 0.45, and so on.

This bound values can determine the existence of too many species (closer bound values), leading to few individuals in each species which allows the progression of low performance individuals across generations, or very few species (further bound values), leading to the competition of individuals with genotyping differences among each other.

The speciation operator in the API executes the Eq. 3.2 in every new individual and assign it to its correspondent specie according to the compatibility value.

3.5.3.3 Selection

The number of offsprings that the system is allowed to generate must be established taking into account that a specified number of elements of each species must be selected. The computation of the amount of offsprings assigned to each species i is represented in the following equations:

$$Offsprings_i = \sum_{j=0}^{N} \frac{Ajusted Fitness_j}{Average Ajusted Fitness} \qquad (3.3)$$

where

$$Ajusted Fitness_j = \frac{Fitness_j}{Size(Species_i)} \qquad (3.4)$$

and

$$Average Ajusted Fitness = \frac{Total Ajusted Fitness}{Size(Population)} \qquad (3.5)$$

After selecting N individuals per species that will generate offsprings, a roulette procedure will be applied to select the mating combinations among the fittest N individuals of that species.

As some rounding to values within these calculations are preformed, the total number of offsprings may be different than the population size. In this case a process called emigration is preformed in which the quantity of excess or missing offsprings are added or taken randomly from a specie.

Moreover, some species may not be able to produce as many offsprings as the previous equations results state that is suppose e.g. if a specie has 2 individuals and the calculations determine that this specie must generate 5 offsprings, this is impossible as each 2 parent individuals can only generate 2 child individuals, making a total of only 4 possible offsprings. For these cases, another process of emigration composed by two steps will take place: (1) a first iteration through all species will adjust their offspring amount to feasible values and collect the amount of emigrants needed to be redistributed; (2) the amount of emigrants will be appended to species that are able to generate more offsprings than the ones the equations settled them to do.

3.5.3.4   Crossover

The crossover process is made within each species and is divided in two steps: (1) a double point crossover in random positions of the parents grid, dividing them in 3 sets of columns and exchanging the middle set between them; (2) a single arithmetic recombination between the weights w_1, w_2, and w_3 of both parents.

The individuals involved in the crossover process as parents are chosen according to their ROI and it is implied that at least half of the offsprings have to be parents. The general rules for possible combinations between parents and children depending on the number of offsprings are the following:

Table 3.1 Examples of crossover operation using the general rules for possible combinations

Offsprings	1 Rule 1	2 Rule 2	3 Rule 3	4 Rule 4 (b)	5 Rule 5 (a)
Parents	●	● ●	● ●	● ●	● ● ●
Children			○ ○	○ ○	○ ○
Offsprings	6 Rule 5 (b)	7 Rule 4 (a)	8 Rule 4 (b)	9 Rule 5 (a)	10 Rule 5 (b)
Parents	● ● ●	● ● ● ●	● ● ● ●	● ● ● ● ●	● ● ● ● ●
Children	○ ○ ○ ○	○ ○ ○ ○	○ ○ ○ ○	○ ○ ○ ○ ○	○ ○ ○ ○ ○

1. 1 offspring—the parent with highest ROI proceed to the next generation.
2. 2 offsprings—the 2 parents with higher ROI proceed to the next generation.
3. 3 offsprings—the 2 parents with higher ROI and 1 of the children resulting from the crossover operation between them proceed to the next generation.
4. X offsprings that imply the existence of an even number of parents:

   a. If X is odd, the $\frac{X}{2}$ parents with higher ROI and the children resulting from the crossover operation between them, except one child, proceed to the next generation.

   b. If X is even, the $\frac{X}{2}$ parents with higher ROI and the children resulting from the crossover operation between them proceed to the next generation.

5. Y offsprings that imply the existence of an odd number of parents:

   a. If Y is odd, the $\frac{Y}{2} + 0.5$ parents with higher ROI and the children resulting from the crossover operation between $\frac{Y}{2} - 0.5$ parents with higher ROI proceed to the next generation.

   b. If Y is even, the $\frac{Y}{2} + 0.5$ parents with higher ROI and the children resulting from the crossover operation between $\frac{Y}{2} - 0.5$ parents with higher ROI, except one child, proceed to the next generation.

For a better understanding on how these rules work with real cases, Table 3.1 contains 10 cases illustrating the relationship between parents and their children in terms of the crossover operation for each one, as well as indicates which of the previous rules applies in each case.

3.5.3.5 Mutation

The mutation operation has a chance P of being triggered and will take place randomly over a number N of columns of the grid. From each of the chosen columns, a number M of cells will also be randomly chosen to be mutated, being their values affected according to a threshold T that will define how big the change is.

These values are constrained as the four conditions for each column stated before in this report must be respected after the operation is completed. Therefore, P is always set to 20%; N can be set between 1 and 2; M is fixed to 2 due to the condition stating that the sum of each column must be 0 (if a cell value is increased, other has to be reduced in the same column); T can take the values 0.1 or 0.2.

The reason of all these constraints relies on the fact that the goal of the mutation operation is to introduce diversity in the population and not slow down the program. After choosing N random columns and 2 different random cells of each, their values will be changed in $\pm T$. All these changes to the grid values are then random, so an auxiliary check function is then triggered to check whether the integrity of the grid regarding the conditions it must fulfill is kept despite the changes. If they are not, the process is repeated for same values N and T, with another random combination of columns and cells and until the auxiliary function checks all four conditions. This is the reason why all the changes involved in the mutation operator must be narrowed into values that do not keep the program stalled for too long.

This operator will also rely on a method called Hyper-Mutation [1], which temporarily increases the mutation rate when a situation of local maximum occurs. This way, when the overall return of the GA does not improve over a predefined number of generations, the mutation rate increases inducing more genetic diversity and the possibility of finding a new better solution.

3.5.3.6   Elitism

Some of the best fitting individuals are eligible to proceed to the next generation without passing through crossover and mutation operations in a process called elitism. The goal is to prevent changing in the performance of the best individuals of each generation, so the fittest individual or set of individuals of the current population, i.e. the ones with highest ROI, do not decrease in the following generation due to crossover and mutation.

For that purpose one of the following conditions must be satisfied: (1) the most 2% fittest individuals of each species must proceed directly to the next generation; (2) if the value of previous condition is smaller than a predefined threshold value (e.g. one), the amount of fittest individuals to have a free pass into the next generation must be set to that value.

In order to track elitist individuals in each generation, the object specie has an attribute consisting of a list of elitists that is updated every generation before the mutation operation. Having this list aside the one containing the current individuals in the population allows to include even the fittest individuals into the process of crossover and mutation, which can open new windows of opportunity to increase the fitness of that species. If by any mean the overall ROI for that species decreases in the next generation, elitist individuals from the current will recover their positions in the current population list.

3.5.3.7 Fitness Function

The main objective of implementing this system is to maximize the return resulting from the investment that was made. So, in order to evaluate each individual of the population, ROI function is going to be applied to pick the best ones for reproduction over each generation of the GA.

Each time the pattern is detected by the algorithm, the program enters in market and only leaves once predefined percentages of profit or loss are achieved. The final ROI is, therefore, also a percentage calculated according to the formula:

$$ROI = \#Wins \times \%TakeProfit - \#Losses \times \%ExitWithProfit \qquad (3.6)$$

where $\#Wins$ and $\#Losses$ represent the number of times the program left the market with profit or loss, respectively.

3.5.3.8 Termination Condition

The termination condition to be applied can affect both the quality and speed of the search, usually in an inverse proportionality relationship. Taking this into account, the program should avoid not only pointless computations, but also immature terminations of the program. Therefore, to fulfill these needs some conditions will be applied to ensure a correct termination of the algorithm:

1. If the algorithm reaches a predefined number of generations;
2. If the highest fitness score for the population does not improve for a predefined number of generations, always smaller than the ones in condition 1.
3. If the highest fitness score for a species population does not improve for a predefined number of generations, always smaller than the ones in condition 1, the fittest individual is kept but that species stops being evaluated.

Even though 1 ensures the program always ends and conditions 2 and 3 solve the problem of pointless computations, the generation numbers that will limit each condition must be carefully chosen in order to avoid the problem of immature terminations. Making sure program execution is properly bounded is key for unnecessary runtime of the system, as well as gives time for GA to improve existing solutions without over-fitting them.

3.5.4 System Optimizations

The main drawback of working with systems that rely on matrix calculations, such as a 10x10 grid of weights, in terms of computational purposes is that the whole process is slowed down by these computations.

Specifically about the FSGW being used in this project some tests were made in a ASUS laptop containing an Intel Core i5 with 2.50 GHz of processor's clock speed, where the average time for evaluating each individual in the GA was about 4 s. For an algorithm run with 25 generations, made over 100 different stocks and containing 100 individuals, the process would take almost 6 days of non-stop computation to finish.

With the goal of reducing this time, rather than heading for a machine with a faster CPU as the first priority, a python multiprocessing module was incorporated in the program. Multiprocessing is a package that supports spawning processes similarly to the threading module, offering local concurrency and the ability to make the most from each available core of the machine. The same test over these new conditions was almost 4 times faster.

Furthermore, the tool Anaconda Accelerate from NVIDIA for taking advantage of Graphics Processing Unit (GPU) of machines with the goal of increase computation speed came to prove unfeasible, as it is designed for common operations that repeat over lots of iterations contrarily to what happens in each iteration of the evaluation operator, which is an highly branched section of code whose behavior depends on the stock being analyzed.

3.6 Chapter Conclusion

This chapter presented an insight on the developed system's architecture. It also contain all particular decisions regarding the API developed for the GA execution, all input parameters' details and possible values, and the way data is represented and manipulated during the whole process, since it is fed by the user to its presentation as output data. The current solution is able to be tested and chapter 4 contains the description of all all preformed tests, their results and descriptive analysis.

References

1. Almeida, B., Neves, R., Nuno, H.: Combining support vector machine with genetic algorithms to optimize investments in forex markets with high leverage. Appl. Soft Comput. **64**, 596–613 (2018)
2. Cervelló-Royo, R., Guijarro, F., Michniuk, K.: Stock market trading rule based on pattern recognition and technical analysis: forecasting the DJIA index with intraday data. Expert Syst. Appl. **42**, 5963–5975 (2015)
3. Fortin, F., De Rainville, F., Gardner, M., Parizeau, M., Gagné, C.: DEAP: evolutionary algorithms made easy. J. Mach. Learn. Res. **13**, 2171–2175 (2012)
4. Leigh, W., Paz, N., Russell, P.: Market timing: a test of a charting heuristic. Econ. Lett. **77**, 55–63 (2002)

Chapter 4
Evaluation

4.1 Tests Scenarios and Objectives

To authenticate the effectiveness of the approach developed a strategy called Back-testing [1] was employed. Backtesting allows the simulation of a trading strategy using historical data to achieve results and evaluate risk and profitability before compromising real capital. If along with it come positive results then the strategy is solid enough and likely to yield profits when implemented. By applying this concept it is possible to test various scenarios without waiting for a long time to obtain results, detect flaws in the approach and analyze where it can be improved.

Over this idea the following three test scenarios were executed:

1. Using the FSGW to detect patterns, described in last chapter, against the TSAG, in order to prove if the new approach overcomes the previous one.
2. Using the FSGW with market exiting parameters (exit with profit and stop loss) that depend on the market volatility over the 3 months prior to the starting day of the test against the same approach using fixed exiting parameters, with the goal of understanding whether this additional measure can improve the results, by relying on it to decide if the risk of leaving a position open for more time expecting a higher return is worth it.
3. Using the FSGW with a speciated GA against the usual GA without species, in a way to verify if the introduction of genetic diversity in the population can act as a key to encounter better individuals in the training phase.

The choice of these scenarios were made taking into account they cover all main features of the system, which, all combined, make it different than existing ones, namely the combination of a GA approach using speciation along with the dynamics involved over the market volatility, and also the FSGW approach never used before.

© The Author(s), under exclusive license to Springer Nature Switzerland AG 2021   43
T. Martins and R. Neves, *Stock Exchange Trading Using Grid Pattern Optimized by A Genetic Algorithm with Speciation*, SpringerBriefs in Computational Intelligence,
https://doi.org/10.1007/978-3-030-76680-1_4

4.2 General Setup

Despite all test scenarios are independent of each other, some parameters and mechanisms are the same throughout the entire evaluation process. Regarding these predefined parameters, Table 4.1 displays their respective values.

The portfolio chosen consist of 100 stocks from the S&P 500 index through a pre-selection process where 503 stocks were trained and the 100 most profitable were chosen as the ones to be analyzed in detail. The choice of this amount was made considering that in a real scenario the management of a bigger portfolio would be impracticable and, despite the optimizations, the algorithm still takes some time when running with a GA with lots of generations. In the choice of this portfolio none fundamental analysis was made, which, if done, could have improved even more the results.

Training and testing periods are treated as percentages because for not all S&P 500 stocks entered the index at the same time, so it is given the same amount of relative time to all stocks, having only the condition that training time percentage must be at lest a whole year, otherwise the training period is not effective enough and the stock is discarded. Despite this, the majority of the stocks have 18 years of historical data available (between January 2000 and February 2018), which leads to approximately 12.5 years of training and 5.5 years of testing, being the point in between both in mid-2012.

The last parameter is related to the FSGW and decides the length of each grid level related to the stock quote. Figure 3.7 serves as an indication for the positioning of the first column of the grid, right in between the second and third cells of the grid and corresponding to the closing price. This Grid Offset indicates how much the stock quote varies between each cell of the grid, e.g., if the closing price for the first column of a specific grid is 15 and the Grid Offset value is 0.5, the price limits of the cell bellow that value are 15 and 14.5, the following one between 14.5 and 14, and so on. Figure 4.1 is an illustration of the given example.

Regarding the evaluation mechanisms to be used in every test scenario some considerations must be taken into account. Firstly, ROI will always be measured in percentages, being the program set to enter in the market when the pattern discovery condition is triggered and will only leave the market once a Take Profit or Stop Loss

Table 4.1 General setup parameters for test scenarios

Parameter	Value
Market	100 stocks from S&P 500
Training period	70%
Testing period	30%
Minimum training time	1 year
Grid offset	0.5

Fig. 4.1 Constraint positioning of the first column of the grid and grid offset

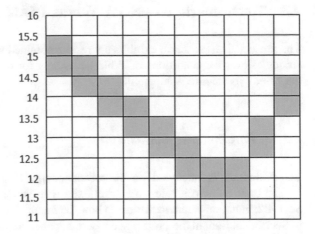

percentage is achieved, being the final ROI for each stock calculated accordingly to the formula 3.6.

Another aspect to take into consideration is related to the fact that for every test all parameter specifications will be the same except the specific one being addressed. For example, in the first test scenario FSGW will be tested against TSAG that uses an adaptable grid and only the closing price for scoring calculation, but all other parameters will be the same. One of the parameters that is always used, except when is specifically indicated otherwise, is the choice of Take Profit and Stop Loss depending on market volatility in the 3 months prior to the beginning of testing phase. Once these volatility values are calculated for every stock, mean and standard deviation of them are obtained and 3 pairs of Take Profit and Stop Loss are defined:

- Lower ones, if the volatility of a specific stock is lower than $Mean - StandardDeviation$;
- Intermediate ones, if the volatility of a specific stock is higher than $Mean - StandardDeviation$ but lower than $Mean + StandardDeviation$;
- Higher ones, if the volatility of a specific stock is higher than $Mean + StandardDeviation$.

Joining all previous, an anaesthetic aspect has also to be pointed for correct understanding of results regarding price window graphs that will be presented. Each can contain three types of vertical lines along them: a black vertical line that indicates a point of entry in the market, a green vertical line that indicates a point of exit from the market with profit and a red vertical line that that indicates a point of exit from the market with loss.

4.3 Test Scenario 1—FSGW Versus TSAG

The presented test scenario exhibits the results obtained when evaluating both these strategies over the conditions of Table 4.1, except for the Grid Offset parameter in the TSAG, as the grid is adaptable.

4.3.1 Parameter Specification

The evolutionary strategy adopts the parameter specification of Table 4.2.

The three pairs in it refer to market exits with volatility influence explained in Sect. 4.2; mutation parameter refers to how the mutation operator works as explained in Sect. 3.5.3, right in the same place where hyper-mutation subject is addressed and it is triggered to double the mutation rate if, after 3 generations, the overall ROI of the population does not increase; species threshold refer to the size of each species regarding their compatibility values as explained in the same subsection as before; and finally when a grid hits the score of 7 units a market entry signal will be given.

4.3.2 Performance Measures

The following graphs show the performance of each strategy regarding the average return during testing period, the average return for every year that the position stays opened in market during testing period and the average return for every year of testing period, whether the position is opened or not in this last case.

Table 4.2 Evolutionary parameters of Test Scenario 1

Parameter	Value
Population	100
Number of generations	25
Generation limit for improval	12
Lower pair (Take Profit %–Stop Loss %)	10–5
Intermediate pair (Take Profit %–Stop Loss %)	15–7.5
Higher pair (Take Profit %–Stop Loss %)	20–10
Mutation (%chance/columns/threshold)	10/1/0.1
HyperMutation (Gens. without improval/increase rate)	3/x2
Species threshold	0.02
Pattern confirmation score threshold	7

Table 4.3 Performance comparison metrics of Test Scenario 1

Parameter	FSGW	Simple
Average market entries	6.17	2.99
Wins (%)	48	50
Losses (%)	52	50
Average days in market	546.3	231.5
Profitable positions	85	80
Non-profitable positions	15	20
Max. profit (%)	120	75
Min. profit (%)	−37.5	−15
Avg. profit (%)	21.3	10.9

Fig. 4.2 Performance comparison graphs of Test Scenario 1

On Table 4.3 each strategy performance is presented according to their entries and days in the market, as well as the profitability of each position.

4.3.3 Discussion of Results and Illustrative Examples

Testing whether the creation of FSGW to contrast with the traditional grid already in use reveals profitable is one of the main challenges of this project. From the graphs presented in Fig. 4.2 we can observe that FSGW clearly surpasses TSAG in terms of average return, except for the average return per year in market, i.e. for each 365 days each position is hold in the market the average return is higher. The main reasons that contribute for this, as observable in Table 4.3, are the fact that the average number of days in market of each stock usin TSAG approach is nearly 2.4 times inferior than in FSGW case, allied to the average number of market entries that are also nearly 2 times inferior and the fact that the percentage of wins is slightly higher. This, however, means that TSAG misses some of the pattern occurrences.

As an example, Figs. 4.3 and 4.4, that display the candlestick representation of Analog Devices Inc. (ADI) stock between late 2015 and early 2017, with market entries and exits for FSGW and TSAG approaches respectively, show an uptrend scenario in which, using a FSGW approach, the pattern was identified three times

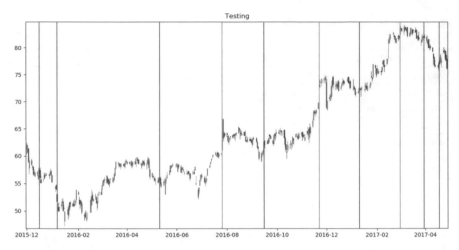

Fig. 4.3 Extract from ADI stock during testing phase of FSGW

Fig. 4.4 Extract from ADI stock during testing phase of TSAG

successfully as the ROI for each entry was positive, but, on the other hand, while using the TSAG approach, none of these patterns were identified.

ADI's stock using FSGW ended up being one of the portfolio's best in testing phase, with a total return of 105%, while using TSAG was only the 47th best with a total return of 7.5%.

Without taking into account any other of previous performance measures, the preceding example could suggest that the FSGW approach was lucky in identifying the pattern as the market was predominantly climbing, specially during the last part of the testing period.

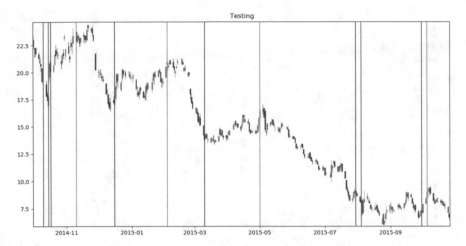

Fig. 4.5 Extract from CHK stock during testing phase of FSGW

To contradict this consideration, Fig. 4.5, that displays the candlestick representation of Chesapeake Energy Corporation (CHK) stock between late 2014 and late
2015, with FSGW market entries and exits, show a downtrend scenario in which,
using a FSGW approach, the pattern was identified three times successfully, as the
ROI for each entry after the highest peak shown was positive and one time unsuc-
cessfully as the ROI for that entry was negative.

On the other hand, Fig. 4.6, that displays the candlestick representation of the
same stock and period of time, with TSAG market entries and exits, only the same
unsuccessful pattern stated in the previous paragraph was identified and it remained
out of the market during the rest of that period, missing at least two windows of
opportunity after the late 2014 peak.

CHK's stock using FSGW ended up being the best of 100 stocks in the testing
phase, with a total return of 120%, while using TSAG ended up only as 43th best
with a total return of 10%.

Both these examples along with already known performance metrics lead to the
conclusion that using TSAG approach in situations of non linear uptrends and down-
trends can lead to false positives or false negatives as an accurate identification cannot
be achieved when it is possible to validate a pattern with a grid that adapts its size to
the price window.

The fact the majority of successful entries in the market occur when the scale of
the grid in a validation situation is similar to the following period in the price window
not only explains why this approach stays much less days in the market, enters the
market much less times, and is less profitable in the testing period compared to
FSGW, but also why the yearly return, referred above in this subsection, is one of
the few metrics that improves with this approach.

Last paragraph observations ended up being some of the main reasons to create
FSGW, as, if we cannot force every price window in the whole market to behave

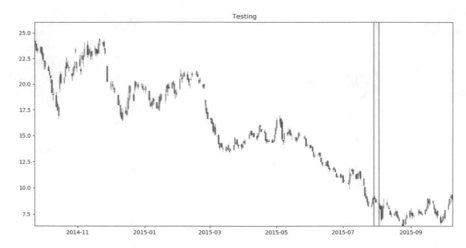

Fig. 4.6 Extract from CHK stock during testing phase of TSAG

similarly in terms of scale like a specific price window, then lets fix a grid within predefined limits in that scale and find the best configuration of scores and weights for a specific market, so all identified patterns are similar within the scale of the grid and not the scale of the market.

Figure 4.7 displays the candlestick representation of First Energy Corp. (FE) stock during the whole testing period, between mid 2012 and early 2018, with market entries and exits for FSGW approach, and serves to illustrate why it was the least profitable with a return of −37.5%. As it is observable, the market is extremely volatile during this period and hence all entries are triggered by a pattern match technique, there are no other factors being taken into account. This is one situation where possibly using fundamental indicators and knowledge/experience of the market would help to blacklist this specific stock from the portfolio avoiding a negative ROI value, but this is beyond the purpose of this work and was not taken into consideration.

The last example of Test Scenario 1 stands as the perfect one to illustrate the main mechanisms of this novel approach in a real case, turning a negative ROI position of the portfolio into a positive one. Figures 4.8 and 4.9, that display the candlestick representation of Pentair PLC (PNR) during the whole testing period, between mid 2012 and early 2018, with market entries and exits for FSGW and TSAG approaches respectively, refer to one of the least profitable of TSAG's portfolio, with a total return of −15%, while for the FSGW approach the total return is 7.5%.

As it is observable the market follows an uptrend in the beginning that is availed twice by the FSGW but never by TSAG, however it becomes extremely volatile at a certain point and TSAG only entered in action here with bad results. All these decisions are explained by the two main aspects that were pointed out to be improved using the FSGW, i.e. the adaptability of the grid and the fact that it only takes into

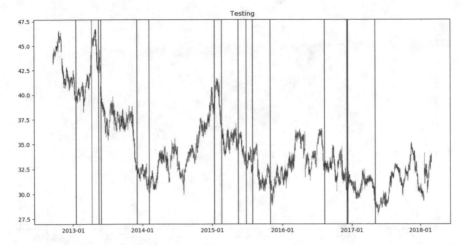

Fig. 4.7 FE stock during testing phase of FSGW

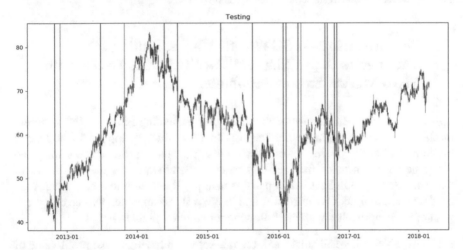

Fig. 4.8 Extract from PNR stock during testing phase of FSGW

account closing price of a stock, leaving some important information given by the candlestick representation regarding the volatility of the price window behind.

In sum, it is safe to say we are debating over two strategies that, for the given portfolio, present positive overall ROIs, but FSGW ended up improving them.

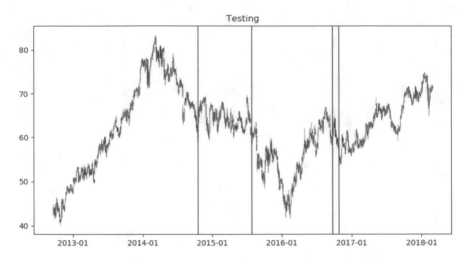

Fig. 4.9 Extract from PNR stock during testing phase of TSAG

4.4 Test Scenario 2—FSGW with Market Exiting Parameters Depending on Volatility Versus FSGW with Fixed Market Exiting Parameters

This test scenario exhibits results obtained when evaluating FSGW in two different modes: with market exiting parameters (exit with profit and stop loss) that depend on market volatility over the 3 months prior to the starting day of the testing and the same approach using fixed exiting parameters, always over the conditions of Table 4.1. As stated in Sect. 4.2, 3 pairs of take profit and stop loss percentages are used while running FSGW normally, and in this test this approach will be compared with other 3 approaches with the following nomination (Table 4.4):

- FSGW 10–5, where all stocks exit the market when having a take profit value of 10% and stop loss of 5%;
- FSGW 15–7.5, where all stocks exit the market when having a take profit value of 15% and stop loss of 7.5%;
- FSGW 20–10, where all stocks exit the market when having a take profit value of 20% and stop loss of 10%;

4.4.1 Parameter Specification

The evolutionary strategy adopts the following parameter specification. Each parameter serves the same purpose as stated in Test Scenario 1.

Table 4.4 Evolutionary parameters of Test Scenario 2

Parameter	Value
Population	100
Number of generations	25
Generation limit for improval	12
Mutation (%chance/columns/threshold)	10/1/0.1
HyperMutation (Gens. without improval/increase rate)	3/x2
Species threshold	0.02
Pattern confirmation score threshold	7

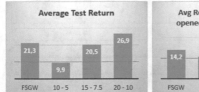  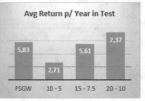

Fig. 4.10 Performance comparison graphs of Test Scenario 2

4.4.2 Performance Measures

The following graphs show the performance of each strategy regarding the average return during testing period, the average return for every year that the position stays opened in market during testing period and the average return for every year of testing period, whether the position is opened or not in this last case.

On Table 4.5 each strategy performance is presented according to their entries and days in the market, as well as the profitability of each position.

4.4.3 Discussion of Results and Illustrative Examples

From the graphs presented in Fig. 4.10 we can observe that, in what concerns the market we are evaluating FSGW, increasing percentages of take profit and stop loss is indeed more profitable. Table 4.5 indicates, as expected, that the higher exiting percentages are, more time opened positions stay in the market, and consequently they enter less times in the market.

The fallout of all this in a market that behaves mostly in an uptrend scenario is that low exiting percentages (e.g. FSGW 10–5) will have lower maximum/average profits and lower losses in terms of its portfolio's worst positions, and on the other hand, high exiting percentages (e.g. FSGW 20–10) will have higher maximum/average profits

Table 4.5 Performance comparison metrics of Test Scenario 2

Parameter	FSGW	10–5%	15–7.5%	20–10%
Average market entries	6.17	7.32	5.91	4.66
Total number of trades	617	732	591	466
Wins (%)	48	42	49	53
Losses (%)	52	58	51	47
Average days in market	546.3	334.6	512.6	691.2
Profitable positions	85	78	83	90
Non-profitable positions	15	22	17	10
Max. profit (%)	120	90	112.5	120
Min. profit (%)	−37.5	−35	−45	−40
Avg. profit (%)	21.3	9.9	20.5	26.9

and higher losses in terms of its portfolio's worst positions, but as the percentage ratio between wins and losses is 2 for 1 respectively, for this type of market the losses being higher does not affect overall significantly.

As it is also observable, the hybrid FSGW that uses all 3 pairs of exiting depending on volatility does not overcome FSGW 20–10, but once again the market in which the scenario is being tested is what allows it to happen, as in a more sideways or sinusoidal market the quantity and severity of non-profitable positions of a FSGW 20–10 approach would reduce its effectiveness. However, even whit this test conditions it is possible to prove this point. Table 4.6 shows all 10 non-profitable positions of FSGW 20–10 compared to the same positions in a portfolio using a normal FSGW approach.

For these positions of FSGW 20–10 to be negative means these stocks' behaviour meet the conditions that reduce the effectiveness of using higher exiting percentages and, as it is observable, using a hybrid approach that reduces exiting percentages revealed a better option in 80% of the cases.

In a more concrete example, Figs. 4.11 and 4.12 that display the candlestick representation of Under Armor Inc. Class A (UAA) along the year of 2016, with market entries and exits for FSGW and FSGW 20–10 approaches respectively, refer to the stock that benefited the most from using a FSGW hybrid approach, instead of FSGW 20–10, with an increase of 25% in total return.

Through observation of the graph itself, it is possible to understand that the stock price is not varying that much over tome, and for each approach a signal for opening a position was given almost at the same time. However, as in FSGW the exiting percentages decided by the volatility of the graph are lower, the approach is capable of taking advantage of slighter uptrends and the position closes with profit right before

Table 4.6 Comparison between FSGW 20–10s worst positions against normal/hybrid FSGW approach

Stock	20–10	Hybrid
ABC	−10	0
ALB	−10	7.5
FAST	−10	15
FTI	−10	−22.5
TGT	−10	7.5
UAA	−10	15
FE	−20	−37.5
MET	−30	−30
HBI	−40	−22.5
SLG	−40	0
Avg. return	−19	−6.75

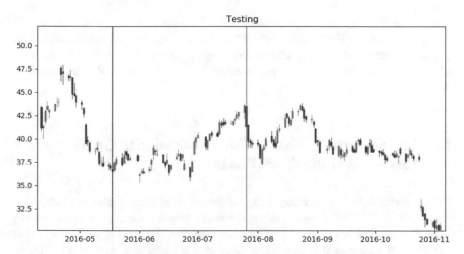

Fig. 4.11 Extract from UAA stock during testing phase of FSGW

the end of July in a small peak. On contrary, FSGW 20–10 approach is not capable of this and if the price does not climb that much once the pattern is discovered, the position will not be closed and takes the risk of ending up in a situation of stop loss. Only this specific case in one of the five years of testing period of UAA ended up being enough for turning a non-profitable stock into a profitable one using FSGW's hybrid approach.

In sum, what can be drawn from this test scenario is that one must know the market in which is putting his money to decide the better exiting configuration. If the market is climbing in overall it was proven that higher exiting percentages bring better results. Once again as all of the scenarios present positive overall ROIs, which

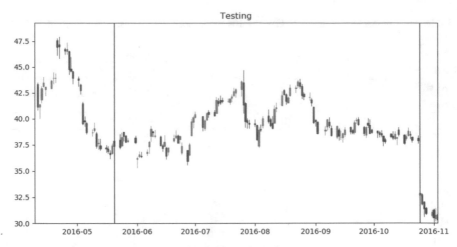

Fig. 4.12 Extract from UAA stock during testing phase of FSGW 20–10

enhances even more FSGW approach, but, as it is known, nothing can be taken from granted in what concerns the stock market, so the approach that benefits the risk/profit relationship and brings along more insurance is the hybrid FSGW with 3 pairs of exiting percentages.

4.5   Test Scenario 3—FSGW Using Speciation in GA Versus FSGW not Using Speciation in GA

The presented test scenario exhibits the results obtained when evaluating both these strategies over the conditions of Table 4.1. As the GA execution occurs on the training phase of the algorithm, beyond the usual testing phase data that is also presented in previous test scenarios it is also being taken into account data from training phase for this scenario.

4.5.1   Parameter Specification

The evolutionary strategy adopts the parameter specification of Table 4.7. Each parameter serves the same purpose as stated in Test Scenario 1.

Table 4.7 Evolutionary parameters of Test Scenario 3

Parameter	Value
Population	100
Number of generations	25
Generation limit for improval	12
Lower pair (Take Profit %–Stop Loss %)	15–7.5
Intermediate pair (Take Profit %–Stop Loss %)	20–10
Higher pair (Take Profit %–Stop Loss %)	25–12.5
Mutation (%chance/columns/threshold)	10/1/0.1
HyperMutation (Gens. without improval/increase rate)	3/x2
Species threshold	0.02
Pattern confirmation score threshold	7

Fig. 4.13 Performance comparison graphs of Test Scenario 3

4.5.2 Performance Measures

The following graphs show the performance of each strategy regarding the average return during testing period, the average return for every year that a position stays opened in market and the average return for every year (whether the position is opened or not in this last case) in both periods.

On Table 4.8 each strategy performance is presented regarding their training and testing periods respectively, according to their entries and days in the market and also the profitability of each position.

4.5.3 Discussion of Results and Illustrative Examples

The use of a speciated approach in this work had the goal of trying to improve the performance of the GA in what concerns the search for a most profitable grid. Observing the graphs of Fig. 4.13 it is possible to realize that overall returns are similar to the point that average return per year in both training and testing phases are separated from each other by less than 1%.

Table 4.8 Performance comparison metrics of training and testing set regarding Test Scenario 3

Parameter	FSGW (Train)	Single (Train)	FSGW (Test)	Single (Test)
Average market entries	18.72	18.65	4.65	4.83
Total number of trades	1872	1865	465	483
Wins (%)	65	68	54	51
Losses (%)	35	32	46	49
Average days in market	1577.6	1586.9	673.3	681.9
Profitable positions	89	89	89	89
Non-profitable positions	11	11	11	11
Max. profit (%)	375	387.5	120	100
Min. profit (%)	60	60	−40	−40
Avg. profit (%)	193.6	200.9	29	26.5

The analysis of the performance metrics presented in Table 4.8 lead us to a similar conclusion, as the most significant differences occur in maximum profit and percentage of market exits with profit metrics of each phases, which are values that, by themselves, do not have enough significance to conclude which approach is the best, as the maximum profit represents just the return of one stock of an entire portfolio and the percentage of wins does not indicate the magnitude of them. It is also visible that the training metric equivalence between both approaches is transposed to the testing scenario.

However, these similarities do not mean that one should discard the speciation approach for three main reasons: (1) the genetic encoding is composed of 57 grid positions and 3 weights whose values can vary, so the number of possible combinations is estimated to be in the order of trillions, which means that even a configuration between individuals and number of generations 1000 times higher than the "100 individuals and maximum of 25 generations" being used would cover less than 0,001% of possible configurations, so indeed the absolute best of these combinations may not be even close of being generated by the algorithm or achieved by genetic operator procedures even using speciation; (2) the weighted approach, the symmetry imposed to the grid that makes mandatory every change to be accompanied by another on the other side of the axis and the range of possible combinations can, all together, lead to a situation where even very different configurations may end up with same results in terms of final return; (3) the cost of using a speciated approach, in terms of time consumption, is roughly similar to not using one, so even if a speciated approach does not ensure that a bigger set of possible combinations is covered, it does cover a bigger area in the map of the possible sets, which is always a more optimistic

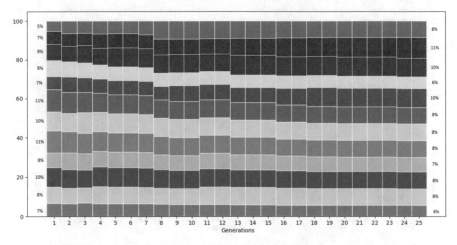

Fig. 4.14 Species profit evolution of SLG stock during testing phase of FSGW

scenario, even if it does not give any guarantees of a better return due to the reasons presented in the two previous points.

The following three figures address examples on how the use of a speciated strategy has improved the overall ROI of the GA in the training phase. As a recall from what was explained in Sect. 3.5.3, the grid weight configuration is what determines in which species each individual ends up.

Figure 4.14 shows the evolution of each species, regarding the profitability of its best individual, during the GA execution of SL Green Realty Corp. (SLG) stock over 25 generations. In this figure are also presented relative percentages related to the ROI of each species in first and last generations for an easier visualization on how each species has evolved.

For SLG's particular case, if this population was not divided into species the best individuals in the first generation were from the "zone" of individuals identified with the colors purple and orange, both having a return of 150% and representing 11% of the best individuals' total returns, meaning that they would likely be dominant over less profitable individuals, such as the ones from the two green species located at each tip of the graph, as well as the red one. However, by letting each species improve their individuals constrained to its scope, the red one ended up being the most profitable with a total return of 190%, and the ones representing 11% of the best individuals' total returns now only account for 8% of it.

Figure 4.15 presents a similar case regarding the GA execution of Republic Services Inc. (RSG) stock over 20 generations, in which species marked in red and orange were two of the best in the first generation, representing 12 and 11%, respectively, of the best individuals' total returns, but ended up decreasing 3% in the final generation. Contrarily, purple and pink species started with a mere quantity of 7% of the total scope and ended up as two of the most profitable with a growth of 4% in their relative profitability over the best individuals of each species.

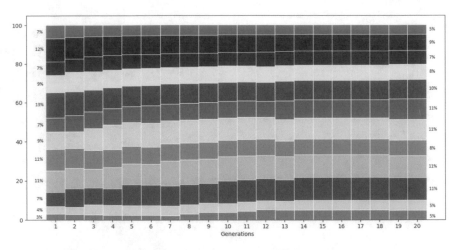

Fig. 4.15 Extract from RSG stock during testing phase of FSGW

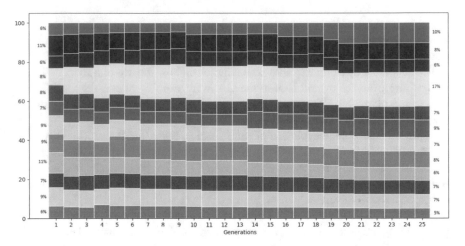

Fig. 4.16 Extract from COG stock during testing phase of FSGW

Finally, Fig. 4.16 regarding the GA execution of Cabot Oil and Gas Corp. (COG) stock over 25 generations is one of the most extreme scenarios where it is visible that a species which was only considered the 6th best in first generation ended up having a profitability that represent 17% of the best individuals' total returns. If the GA was not using speciation, this individual would eventually be crossed with a random one in the population instead of only ones with the same genetic compatibility, condition that was crucial to the creation of an extreme profitable individual.

In sum, it is proven that for training a FSGW using a GA speciated approach is possible to achieve at least as good results as a non-speciated approach, however there were presented cases where the use of species allowed a turnover in the profitability of individuals that were weaker in early generations, and as this approach is not more time consuming than using the regular GA, it is definitely to be considered. Also due to the huge quantity of different evolutionary parameter configuration more ways of trying to improve the speciated could have been addressed, but once again the main purpose of this system was not that.

4.6 Overall Analysis

This section serves as a summary of the results presented in previous test scenarios regarding all different approaches presented, where some overall comparisons and observations will be made. Recalling all different approaches being tested we have the main FSGW and all its variants (exit pairs 10–5%, 15–7.5% and 20–10%, as well as the approach with a single species) and also TSAG.

Before heading to the main subject of this sub-section, it is worth to mention that while testing each of previous approaches it took approximately 1 day and 2 h of non-stopping computation per execution. For that reason there was not enough time to execute sufficient tests so that an average run of each approach could be presented. Nevertheless, as FSGW is the main system being tested, a series of 10 runs for that approach were made and Table 4.9 show the comparison of the run chosen for the previous test scenarios with the average results of the 10 runs, also including the highest and lowest result for each parameter.

It is observable that the run chosen for the testing scenarios stays close to average values and even the disparity between highest and lowest result for each parameter is narrow, which leads to the conclusion that the chosen run of the algorithm can be accounted as well representative of what FSGW is able to achieve, as well as the approach is reliable since it always leads to similar results.

Heading to the overall analysis, Fig. 4.17 joins all the different tested approaches together in a graph that shows the evolution of their average ROIs over the entire testing period. The lines highlighted with red and green glows are FSGW and TSAG approaches, respectively, and this serves not only to show the magnitude of the improvement that using a weighted grid with a fixed size approach over the traditional existing approach has, but also to show that from 5 FSGW approaches the only that does not overcome TSAG is the one in which the market exiting strategy is the most conservative one.

Pursuing now to an yearly analysis of ROI's evolution within the studied approaches, Fig. 4.18 presents another clear prof of the FSGW approach dominance over TSAG in terms of average ROI. Excluding 2012 and 2018, which only account for roughly 2 month in the testing period each, only 2015s average ROIs for FSGW's approaches bumped more than TSAG. Yet, the overall yearly returns never drop more than −2,5% while in the most positive scenarios the traditional approach is always at least 2% less profitable than the main FSGW, and in general less profitable than at least 4 out of 5 FSGW variants.

Table 4.9 Comparison between FSGW run chosen for test scenarios against the average of 10 executed runs

Parameter	FSGW	FSGW Avg
Average market entries	6.17	6.12 [5.89; 6.29]
Wins (%)	48	48 [47; 49]
Losses (%)	52	52 [51; 53]
Average days in market	546.3	556 [553.7; 572.1]
Profitable positions	85	84 [82; 88]
Non-profitable positions	15	16 [12; 18]
Max. profit (%)	120	117 [105; 127.5]
Min. profit (%)	−37.5	−37.5 [−45; −27.5]
Avg. profit (%)	21.3	20.7 [19.5; 22.4]

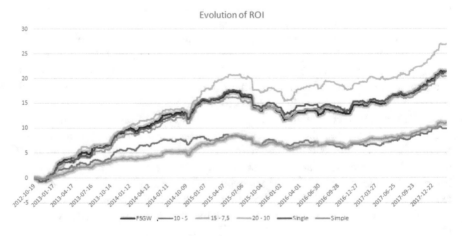

Fig. 4.17 Average ROI evolution over the entire testing period

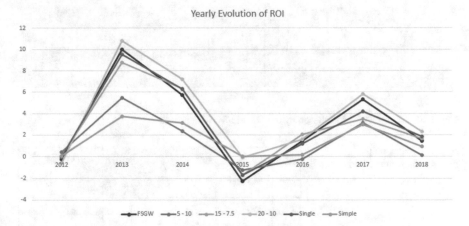

Fig. 4.18 Average yearly ROI evolution over the entire testing period

Fig. 4.19 Percentage of
win/loss results from FSGW
approach and its variants

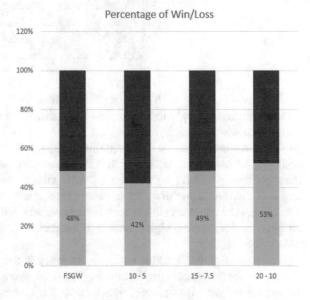

Another important detail important to point out is related to the duality between the percentage of exiting positions that won and lost. Watching out on the graphs presented in Fig. 4.19, at first sight it might appear that FSGW strategy and all its variants are not that good, as the duality referred above tends more to higher cases with losing positions than winning ones.

However, and as it has been said before in this report, the relationship between a win and a loss is always from 2 to 1, i.e., once a market exit strategy is defined it is always assured that, for every win, it is possible to have two losses and yet not being in a negative balance. Addressing the worst scenario in the following graphs, even with only 42% of wins against 58% of losses, if the total number of market exits

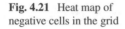

Stock: UAA   Stock: FE   Stock: CHK

Fig. 4.20 Example of grids chosen by the GA in FSGW approach

Fig. 4.21 Heat map of
negative cells in the grid

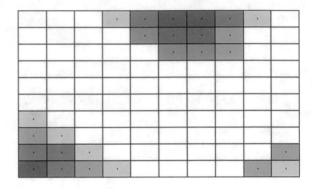

totals 100 will end up with 84% of profit against 58% of loss. With this win-loss
relationship settled, only in a scenario of less than 33% of wins the balance would
start to become negative.

One last remark must be done regarding the overall results of these test scenarios
and is related to the grids chosen by the GA to proceed from trading to testing phase of
the algorithm. All Bull Flag grids have 2 main visible characteristics: the downtrend
and consolidation zones of a perfect pattern market with the highest scores and the 3
most penalized zones located in both bottom corners and on top of the 5th, 6th, 7th
an 8th columns where a high presence of the market signal is a huge indicator of a
failed identification.

One particular detail stands out in the majority of the resulting grids of FSGW's
GA procedure as shown in the examples of Fig. 4.20, in which, from the 3 penalization
zones, the bottom right corner, right bellow the highest scoring positions of the
consolidation zone of the pattern, is not that much punished with bad scores as the
others. In fact if we treat the grid as a two dimensional referential centered in the
middle of it, the whole 4th quadrant is the less penalized area of the entire grid.

Going even further on analyzing the way this penalization zones interfere within
the market signal, if a "heat map" of the common negative cells of the grid was
designed, it would look something like Fig. 4.21.

The explanation stands as a relationship between the location of the penalization
zone with the way the signal is supposed to behave in a Bull Flag pattern. A market
signal that enters the bottom left penalization zone is likely to proceed into a steep
downtrend and even if it recovers into what remains of the desired pattern, it won't

look like the template. Following the same line of thought, a market signal that enters the top penalization zone is more an indicator of a sideways scenario than a Bull Flag, contradicting also the shape of the correct template. On the other hand, if a signal reaches the bottom right penalization zone with a good total score already from the previous positions, this means that it has behaved correctly more than half of the grid. Adding to this, two more reasons arise to allow the signal to be in that position without penalizing it too much: the consolidation phase may be a late one, or, even if it is not, confirming a pattern that ends in the lower zones of the grid is always safer than ending in the highest, as it has more room to climb if it indeed starts climbing as expected.

4.7 Chapter Conclusion

Within this chapter were presented all test scenarios that were made using a portfolio of 100 stocks from S&P 500, over training and testing periods representing 70 and 30% of each stock's available data, which, in the majority of cases, consists on 12.5 years of training (between early 2000 and mid 2012) and 5.5 years of testing (between mid 2012 and beginning of 2018). The designed algorithm, using a fixed size grid and a new way of attributing scores, beat the existing adaptable and simple grid, having twice the average return on investment over the testing period. It was also proved that using exiting percentages based on market volatility avoids big losses in low volatile/sideways markets, as well as using speciation guarantees that the range of possible individuals to be trained by the GA is wide, even though that optimization did not improved significantly the overall return on investment.

Reference

1. Ni, J., Zhang, C.: An Efficient Implementation of the Backtesting of Trading Strategies, pp. 126–131. Springer, Berlin (2005)

Chapter 5
Conclusions and Future Work

5.1 Summary and Achievements

The presented work proposes a portfolio management system using a GA with speciation along with a new approach to pattern identification of the template based approach using a grid of weights.

To validate the implemented strategy it was compared against the already existing that using a grid of weights; against itself using different input parameters regarding the way market volatility affect the program; and again against itself using the same approach but without species. The period of training and test were 70% and 30% respectively, applied over the available data of a portfolio containing 100 stocks from S&P 500, which, in the majority of the cases stand for 18 years of data between January 2000 to the first semester of 2018, that can be divided into, approximately, 12.5 years of training and 5.5 years of testing, being the point in between both in mid-2012.

The developed algorithm is able to simulate all test scenarios and the quantity of parameters it gives for user to choose, specially the ones referring to the evolutionary algorithm, opens up the possibility of many to be tested depending on the needs of the user.

Achieved results of the proposed strategy compared with the one already existing under the same parameter specification were very promising, specially regarding the overall returns over the testing period. Various variants of the proposed strategy were also tested in order to draw some conclusions about how results react to changes in some parameters, yet the majority of them still beat the traditional approach.

5.2 Future Work

In this sub-section, several limitations of the algorithm are addressed, along with some improvements that could repair those limitations as well as the overall functioning of the system.

© The Author(s), under exclusive license to Springer Nature Switzerland AG 2021   67
T. Martins and R. Neves, *Stock Exchange Trading Using Grid Pattern Optimized
by A Genetic Algorithm with Speciation*, SpringerBriefs in Computational Intelligence,
https://doi.org/10.1007/978-3-030-76680-1_5

- The algorithm is not yet prepared to run in real time and be constantly updating information, analyzing it and notifying the user of what to do next, which would clearly be the next step as the validity of the system is completed;
- Extend the testing scenario from S&P 500 to other markets and understand how the developed approach behaves within new scenarios;
- Increase new patterns to the scope of FSGW analysis and allow the short selling mechanism to increase the time in the market and possible winnings;
- Allow a multi-pattern detection approach, in which various patterns would be analyzed at the same time and decide what to do depending on relative scores of each one;
- No kind of fundamental analysis or risk measurement is applied in order to decide if at some point the user wants a position closed because the stock is considered too risky, even if the algorithm says otherwise. For that reason we should enable the possibility for it to be removed from the portfolio, even if technical indicators are good;
- During testing phase (or in case of it being used in real time) there are no automatic measurements being done simultaneously, e.g. tracking volatility and moving averages, so that some parameters could be adjusted in real time depending on this analysis with the goal of reducing risk and increasing the profit;
- The amount of input parameters are so wide that some kind of information regarding the current configuration could be given, like creating a cache of already tested scenarios and use it to inform the user in future tests what happened when some subsets of parameters were configured in the exact same way in the past;

Printed in the United States
by Baker & Taylor Publisher Services